Living with Tiny Aliens

# gROUNDWORKS|

## ECOLOGICAL ISSUES IN PHILOSOPHY AND THEOLOGY

**Forrest Clingerman and Brian Treanor,** *series editors*

# Living with Tiny Aliens

## The Image of God for the Anthropocene

*Adam Pryor*

**Fordham University Press** | *New York 2020*

Visit us online at www.fordhampress.com.

Library of Congress Control Number: 2020902553

Printed in the United States of America
22 21 20   5 4 3 2 1
First edition

*For Henry, Greta, and Linus*

# Contents

Introduction: Being in Outer Space — 1

1   Exoplanets and Icy Moons and Mars, Oh My! — 15

2   Astrobiology's Intra-Active Aliens — 32

3   Being a Living-System — 46

4   The *imago Dei* as a Refractive Symbol — 62

5   Conceptualizing Nature — 86

6   The Anthropocene as Planetarity in Deep Time — 104

7   An Artful Planet — 128

8   Living-Into Presence, Wonder, and Play — 146

Epilogue: *Ad Astra Per Aspera* — 188

*Acknowledgments* — 201

*Notes* — 203

*Bibliography* — 241

*Index* — 263

# Living with Tiny Aliens

# Introduction:
# Being in Outer Space

The technologies that put human beings into outer space are amazing. It is astonishing that we can exist in the midst of the foreboding vacuum of that inky blackness. For instance, consider the Manned Maneuvering Unit (MMU). This device fit over the life-support backpack of the baseline Extravehicular Mobility Unit (EMU). The MMU propelled an astronaut by releasing gaseous nitrogen from twenty-four different thrusters. However, after safety reviews conducted following the *Challenger* disaster, the MMU was quickly retired (it was used on only three missions in 1984), but the freedom and adventure inspired by the MMU—allowing an individual to fly through space—are hard to shake. It even imaginatively appears in the 2013 film *Gravity* to jet George Clooney from place to place.

While the possibilities of the MMU tap into the dreamy imaginings of science-fictional aspirations, the EMU remains a marvel of engineering—it really is an individual-sized microhabitat. It can support an astronaut for eight-and-a-half hours, providing for all the essential needs of the human being in space and even some additional comforts (for example, EMUs on the International Space Station include glove heaters for when astronauts are in the shadow of the Earth).

To get a glimpse of the engineering complexity of the EMU, consider the fourteen layers that make up the suit. The first three layers consist of the Liquid Cooling-and-Ventilation Garment (LCVG). The LCVG includes more than ninety meters of plastic tubing laced into a spandex suit. Connecting to the Primary Life-Support System, the LCVG's tubing and ducting help control body temperature (running cooled water through its tubes), ventilate oxygen and carbon dioxide (pulling in air at the wrists and ankles while putting out pure oxygen through a duct at the back of the helmet), and manage perspiration (recycling into the water-cooling system).

On top of the LCVG are the three parts of the bladder layer (consisting of nylons, polyesters, and neoprene) that control pressure and hold in oxygen. Holding in oxygen is obviously important, but pressure control is an equally critical feature. Exposure to the pressure vacuum of space would cause the air inside an astronaut's body to expand and the liquids inside the body would essentially boil (because gasses escape from liquids more easily at lower pressures). The bladder layer has to function like a balloon to simulate atmospheric pressure for the wearer.

The next seven layers of the suit are forms of Mylar insulation intended to regulate temperature—like a giant thermos. The drop in pressure and gas molecules in space means that there are *severe* changes in temperature based on solar radiation, ranging from 120 degrees Celsius (248 degrees Fahrenheit) when in view of the sun to –100 degrees Celsius (–148 degrees Fahrenheit) when shaded. Finally, the outermost layer is a blend of fabrics including now familiar, high-tech materials like Gore-Tex (for waterproofing), Kevlar (as used in bulletproof vests), and Nomex (a flame-retardant) for durability and protection from small, high-speed objects that could be encountered in space.[1]

Gas, liquid, temperature, and pressure management all make the EMU into a virtual habitat. The suit extends the place of human being into otherwise unlivable contexts in the universe. It is a precarious and short-term solution as far as habitability goes, though: like wrapping a human being into a complex balloon situated inside a well-fitting thermos that can take care of our most primordial needs for a third of a day.

Appreciating the precarity of living in an EMU should make us marvel at the Earth. We may not always feel the same sense of astonishment, but the Earth parallels the EMU in a critical sense. The Earth also provides human beings a place to live in the otherwise deadly vacuum of space, though on a different scale both temporally and spatially.

Thinking of the Earth this way always brings to mind, for me, photographs of our planet taken from space—especially the image from the *Galileo* spacecraft. It was taken as *Galileo* was passing Earth on its mission to Jupiter in 1990, and data from that flyby was used to suggest that biosignatures could be detected from space. Because *Galileo*'s instrumentation was—at its launch time—typical for any planetary probe NASA might send out, Carl Sagan and his team wondered if one would be able to recognize that there was life on Earth from the data collected *if one were not specifically looking for life*. They looked, specifically, for a "marked departure from thermodynamic equilibrium."[2] Such a disequilibrium would provide a necessary, though not sufficient, indication of the presence of life if one could eliminate all abiotic means of accounting for it.

Their findings were surprising. The *Galileo* flyby indicated just how difficult it would be to discover an advanced technological civilization. The spacecraft took pictures of the Earth's surface at the highest resolution (1–2 kilometers per pixel), but it was only with *a posteriori* knowledge that these images served as definitive evidence for a technological civilization. The principal problems were how much of a planet would need to be imaged and what level of resolution was needed to discover an advanced technological civilization.[3] As the authors' conclusion starkly notes, "Most of the evidence uncovered by Galileo would have been discovered by a similar fly-by spacecraft as long ago as about 2 billion ($10^9$) years."[4]

However, one could hypothesize the presence of a water-based biotic system from the data gathered (using information derived largely from the ultraviolet and near-infrared mapping spectrometers) that would photodissociate water, producing otherwise unattainable levels of atmospheric oxygen along with levels of methane well out of equilibrium with the oxygen-rich atmosphere. Yet, "[H]ow plausible a world covered with carbon-fixing photosynthetic organisms, using $H_2O$ as the electron donor and generating a massive (and poisonous) $O_2$ atmosphere, might be to observers from a very different world is an open question."[5]

Whether one is considering habitable conditions as vast as a planet or as small as the EMU, maintaining certain basic conditions is essential for living things. A living thing and its habitat (if viewed in isolation from each other) seem strangely well designed for each other. In short, the account one gives of either a living thing or a habitat is simply insufficient if we ignore this connectedness. To understand a living thing, truly, requires understanding its habitat.

Astrobiology recognizes this deep-seated attunement, but that is hardly new. It stands in connection to wider intellectual histories, albeit at times less emphasized, that pay attention to the correlation of living beings with the wider world. Western monotheists can look to the cosmogonies of the Hebrew Bible that integrate attunement to the very grammar of creation. The thoroughgoing modernist can point to renewed interest in the study of Alexander von Humboldt and Maine de Biran. Or, the biologist can find nuanced accounts of attunement in the work of Jakob Johann von Uexküll (and perhaps even earlier biologists).

Yet my own sympathies for astrobiology remain, perhaps as an expression of the theological context from which my work proceeds—a context that takes the idea of attunement quite seriously. As the reader will certainly discern, I am indebted to a tradition of neo-platonic thinking in Christianity running ". . . through the Greek Fathers, the Pseudo-Dionysius, Anselm, the Victorines, Alexander of Hales, and Bonaventure. It continues in subsequent thought in Nicholas of Cusa and the Platonists of the Renaissance, surfacing again in German romanticism and reappearing in various forms in the twentieth century."[6] Studying our sense of "bodily integrity and harmonious attunement to our environment entails and explains our sense of ultimate concern."[7]

In a sense, all I am suggesting is that astrobiology is part of a long tradition studying the meaningful correlation of life with the environment. Still, this in itself is important because it indicates that *astrobiology is not merely about finding aliens*—whether intelligent or microbial. Astrobiology is far more concerned with understanding how the relationship between systems of living beings and the environment unfolds; astrobiology wants to discover how it is that life manages to appear in the universe. As Carl Pilcher—the former director of the NASA Astrobiology Institute—puts it, astrobiology is "a quest to understand the potential of the universe to harbor life beyond Earth."[8] While finding aliens would *confirm* this potential, astrobiology itself is primarily concerned with *understanding* this potential.

Perhaps recognizing and taking seriously this narrowed scope of astrobiology is one of the most difficult aspects of communicating the meaning of this field to wider publics. We want to rush headlong toward sentient aliens that are curiously like ourselves. Fueled by our science-fiction dreams and nightmares—as with *Arrival, Contact, Independence Day, Solaris,* or *The Sparrow*—we leap ahead: Astrobiology becomes the field studying first contact with recognizably intelligent (if only sometimes benevolent) lifeforms. In reality, astrobiology—as a scientific endeavor—is mostly doing something much more modest and potentially far more important.

Staying with this more properly scientific side of astrobiology plunges us into the vastness of a cosmos that quickly becomes unimaginably capacious. The places where this potential for harboring life arises may be more common than we thought, but these harbors arise over *vast* distances both spatially and temporally. As astrobiologically aware human beings, we must confront our deepened anxiety arising in the face of our own contingency—realizing how deeply tethered we are to the moments this pale blue dot exists in the universe. At the same time, our astrobiological awareness is opening a horizon to the exciting possibility of understanding our humanity in relation to not only a planet burgeoning with life but also a cosmos pregnant with living possibilities.

## Finding the Meaning of Being Human in Space

Touching upon both the deep connectedness between living-systems and habitability and the cosmic context that shapes the understanding of this connectedness at work in astrobiology, this work examines how a new context for conceptualizing the ways in which human beings understand their belonging in and to the world is emerging. Broadly, what follows is an effort to imagine how an individual's meaningful existence persists when we are planetary creatures situated in deep time—inextricably conditioned by the nonseparable relationality of living-systems and habitable environments. This is an approach to astrobiological humanities: considering how astrobiology helps figure expressive modes by which human beings process experience.

At this level, the project should be of interest to anyone who wants to trouble the various ways we circumscribe human being in a dermal metaphysics: parceling apart the subjectivity of an individuated self and the objectivity of a world at the boundary of the skin. Astrobiologically, we must situate the sense of "self" in a cosmic scale that is astonishing and threatens to eclipse the individual human being. Can we adopt alternative primordial units for ontology that might be more consistent with the astrobiological perspective? What would be the effect of such a shift for personal responsibility that has traditionally framed our accounts of the modern, liberal individual's motive power? What ontological language would we need to engage in such reflection?

Rethinking how to conceptualize human being in its planetarity creates a specific challenge for any form of theological thinking

rooted in a strong sense of individuality. Herein, I understand theological thinking not as a mere apologetic exercise; instead, theology
is an act of interpreting symbols that help meaningfully order our
existence. As such, the "theology" pursued here is a form of philosophical or correlational theology rooted in understanding religions
as cultural phenomena. It is a process through which we intentionally engage with all sorts of planetary others and orient ourselves
in the midst of the world in accordance with religious symbols that
facilitate meaningful living.

My hope is that adopting this understanding of religions as cultural phenomena does not entail abandoning the potential significance of *doctrine*. Too easily, doctrines become historic arbiters of
right and wrong interpretations of faith with merely *internal* significance. When this happens, a doctrine merely formalizes foundational propositions relevant within a given community: They *inter-
nally* help interpret and clarify the meaning of principles uniting a
religious community. Externally, their significance is only in giving
insight into the learned expressions of the beliefs deeply held by a
particular community or in solidifying the continued dominance of
subject-positions privileged by this doctrinal discourse.[9] In any case,
the significance of doctrine remains strictly tied to the particular
community of faith.

My aim is to cut against the tendency to make theology into
apologetic defense by thinking about the significance of doctrine as
a particular type of *public theology*. This is especially important for
projects in theology and science that can appear to be extremely
apologetic: merely ensuring that specific religious doctrines and
metaphysical schemas survive cultural shifts and the growth of
scientific knowledge. Later, I will clarify what "public theology"
entails. For now, I simply contend that it denotes a theology intentionally fostering a wider societal set of goods extending beyond the
confessional proclivities of a given community. It must commit to
interpreting sets of religious symbols—however they might appear
or develop—that when enacted by human beings establish powerful
moods and motivations as a means to ordering existence in congruence with ultimacy.[10]

For the traditional doctrinal theologian, the book extends a well-
established line of thinking opened by liberation and constructive
theologians. For constructive theologians, though, I hope the book
is a cautionary tale: encouraging us not simply to throw out doctrine
as irrelevant to or anachronistic for contemporary theological pur-

suits. By focusing on the significance of astrobiologically reframing a single doctrine—the *imago Dei*—that has often been cast off as anthropocentric, sexist, and oppressive, I hope to offer an exemplar for the wider methodological significance of constructive theologians reclaiming doctrine.

If one bears this methodological issue in mind, the significance of the book is threefold. First, it is an in-principle argument with interfaith consequences. The account of theological thinking offered here is flexible enough that others could apply it to sets of symbols in *various* religious traditions. If changing the context of engagement—in this case from a principle of human individuality to human belonging understood astrobiologically—for interpreting a symbol shifts how it meaningfully orders existence, then theologians and religious practitioners would need to consider the identification of new differences or accords that exist across the interpretation of symbol sets from various religious traditions *given a fundamental shift in the sense of human beings' belonging to the world*. Though not my primary focus, such comparison is an in-principle possibility.

Second, the book addresses the Christian tradition directly and the outsized importance given to the *imago Dei* (given the paucity of Biblical references to the concept) as a defense against the overwhelming existential anxiety produced by the contingency of creation. We need not be; nonetheless, we are. The *imago Dei* helps explain the force of that "nonetheless": justifying our existence despite its being neither necessary nor impossible. As such, the doctrine works a bit like a backstop in baseball: Contingency may roll along so far, but it cannot pass the assertion that our image and likeness to the divine make us irreplaceably *purposeful*.

Various theologians have imagined well what it means to face our personal contingency in the sweeping midst of society: imagining the world humming along without *me* being a part of it. An existential theology could follow this track: simply assuming astrobiology does not present a fundamentally new challenge to our confrontation with contingency. Instead, the problem would be a scalar one. How do we affirm our acceptance by God in the face of an ever-widening cosmos that threatens to make us irrelevant?[11] However, such an approach ignores the prospect of astrobiology's re-imagining our way of being-in-the-world, a way of being-in-the-world that cannot be separated from our belonging-to-the-world—a way that makes the intra-action of self and world primary and always situates the self in the wider context of participation in a living-system. In this shifted context, the typical interpretations of what it means to be the image of God, founded on individuality, should be re-imagined.

We have only recently begun, theologically speaking, to think about the contingency of our species in terms of this sort of planetarity as it contrasts a pervading individuality, particularly with regard to its astrobiological significance. My work represents another step in that direction: considering how we construct a theological anthropology—an existential theology that takes our individuality seriously even if it is not ontologically primordial—in the face of catastrophic change.[12] Certainly this work is of interest to those studying theology and science (as a constructive account of doctrinal theology articulated in terms of astrobiology), but it should be significant to other theologians as well. The constructive account of the *imago Dei* offered here as a "refraction" or "symbol" to be lived-into critiques Christian theological anthropologies founded on the individuality of the human person and proffers a vision of the *imago Dei* for a more capacious anthropology.

Third, my approach to theological thinking is explicitly not dogmatic. My goal is not to interpret the doctrine of the *imago Dei* as relevant within the confines of a particular confessional community. Certainly, one could understand what follows in a purely confessional vein, but my hope is that it may do more as well because symbols have a deep, persistent resonance not easily made mute. Edward Farley describes quite well what I intend to pursue.

> In the sphere of the interhuman, human beings relate to each other, not merely as functionaries in a preprogrammed bureaucracy, but in mutual perceptions of their vulnerability, needs, pathos, possibilities, and mystery. In the sphere of relation human beings continue to experience mutual obligation, guilt, and resentment, gratitude, limitations on their autonomy, and mutual activities of creativity. From such relations are born notions of personhood, justice, mutual obligation, and even truth and reality. *When a society or individual presupposes a god-term as something normative, something to appeal to, it is not simply appealing to the symbol, for the symbol has brought to expression a deeper normativity at work in the sphere of relation.* Here we have the primary reason for thinking that the words of power are not utterly extinguished. That which makes its appeal through them, the enchanted myster-

ies of human beings together in relation, has not been totally abolished.[13]

Symbols are not the strict property of confessional communities; they exist in the wider culture. A person far from a symbol's originating community can take it up in various ways. The *imago Dei* is just such a symbol: one wandering from its origins. It is a symbol with public resonance—conveying human dignity and distinctiveness—that can meaningfully orient the existence of persons both inside *and outside* the symbol's community of origin. When it is understood as such, to invoke the *imago Dei* is to constitute the bounds of meaningful human being in the sphere of our interhuman relationships.

Thus, this work can serve as a theological project that might be relevant for *anyone* who finds the *imago Dei* to be a means of meaningfully orienting her existence. I assume that this symbol can persist in meaningfully orienting our existence even when divorced from its originating religious context. So while I will reflect deeply on the trinitarian and christological elements that have traditionally shaped how the *imago Dei* is theologically interpreted, I do this to enrich our understanding of the ways in which this symbol has engaged with ultimacy and can continue to communicate how it is of concern even outside explicitly religious communities.

## A Transdisciplinary Project

In apologetics or dogmatic theology, we could be quite clear about the scope of a project like this one. However, a broad, public theological project does not fit neatly into a traditional "field" of study with rigorous methods and objects of investigation. What group of scholars has the authority to reflect on the *imago Dei* in this way? Given a waxing suspicion in the academy toward the efficacy of strict disciplinary approaches[14] and my use of astrobiology (which itself seems to be composed of many traditionally understood fields), the scope of this project becomes quite unwieldy.

I do not wish to run too far down a methodological path, but, at the outset, I hope to avoid some confusion that might arise from using four contested terms: cross-disciplinary, multidisciplinary, interdisciplinary, and transdisciplinary. I will follow a widely employed schema initially developed by P. L. Rosenfield for organizing different types of cross-disciplinary scientific inquiry.[15] "Cross-disciplinary" is the broadest term, referring to any research that makes use of more than one discipline. Intentionally generic, it will be used when it is unclear—or it has yet to be clarified—how a more specific approach (multidisciplinary, interdisciplinary, or transdisciplinary) is being employed.

Multidisciplinary, interdisciplinary, and transdisciplinary specify forms of cross-disciplinary engagement. There is no clear consensus on what multidisciplinary, interdisciplinary, and transdisciplinary research specifically entail in distinction to one another. It is generally assumed that from multi- to inter- to transdisciplinary inquiry there is an increasing interrelation between the fields in question. The problem being pursued increasingly fuses the various disciplinary approaches so that we can no longer rend them apart as the knowledge produced increasingly relies on the assumptions of another discipline in the inquiry.[16]

Multidisciplinary research, as the simplest form of cross-disciplinary engagement, is characterized by methodological parallelism. It has a serial or sequential quality to it—placing one discipline's findings next to a different discipline's findings such that they support or constrain one another. Interdisciplinary research represents tighter cross-disciplinary engagement than multidisciplinarity, but less engagement than transdisciplinarity. It involves cross-disciplinary engagement that feeds back into the participating monodisciplinary knowledge sets. This creates closer coordination between the research agendas of the participating disciplines, "yet the participants remain anchored in their respective disciplinary models and methodologies."[17]

Transdisciplinary research is highly integrated and transcends any single disciplinary structure. The participants in transdisciplinary research work with a shared conceptual framework that extends concepts, theories, or methods from each discipline in support of a collaborative working relationship. A. K. Giri suggests that in transdisciplinary research, those participating understand themselves to be part of a team, not a mere collective, that requires commitment to the various points of view involved, despite how the findings of another field might overturn critical insights from one's primary discipline: requiring a widening of one's disciplinary horizons.[18] In transdisciplinary research, one must abandon unwavering commitments to monodisciplinary constructs because of an orientation toward resolving identified social problems. There is a praxis element to transdisciplinary research, and this facilitates the rejection of any fragmentation of knowledge. Transdisciplinary engagements seek a way of understanding that does more than consistently hold together complex and divergent ways of knowing. These engagements do this in pursuit of a sense of social well-being that we cannot adequately characterize with a fragmented knowledge base.[19]

Transdisciplinary engagement does not, however, specifically aim at forming a new, more abstract field of study with a formalized method. Instead, it is indicative of a certain awareness on the part of the transdisciplinary researcher that her own disciplinary boundaries can be limiting for engaging a given object of study or common interest. To pursue the sense of social-wellbeing that motivates transdisciplinary engagement, one must commit to a degree of disciplinary skepticism that gives space for other fields of study to enrich our accounts of these common concepts.[20]

No single one of these forms of cross-disciplinary engagement is inherently superior to the others. Each can serve important, *different* purposes in a given cross-disciplinary engagement. Our tendency—once we have established a distinction between these three approaches—might be to simply ask, "Is astrobiology multidisciplinary, interdisciplinary, or transdisciplinary?" or "Is theology and science research multidisciplinary, interdisciplinary, or transdisciplinary?" There is good reason to resist the impulse toward this sort of categorizing.

We might take astrobiology as a ready example, as it seems to employ each of these modes of cross-disciplinary engagement to different effects. For instance, in thinking about a definition of life, astrobiologists may engage in multidisciplinary reflection that is concerned with how findings related to stochastic processes can constrain thinking in systems biology that in turn could affect how one gives an account of a "population" in astrobiology. In this case, a more robust cross-disciplinary engagement might confuse what is at stake.

Alternatively, NASA's astrobiology research is guided by three overarching questions: How does life begin and evolve? Does life exist elsewhere in the Universe? What is the future of life on Earth and beyond?[21] Even tentatively venturing answers to these three questions requires an integration of research agendas that goes well beyond mere multidisciplinarity to some sense of interdisciplinarity.

Finally, an astrobiologist like David Grinspoon is taking his readers well into the realm of transdisciplinarity by linking astrobiology to preserving human social-wellbeing related to planetary sustainability. To make matters more complicated, however, in making his transdisciplinary argument, he is clearly making use of multidisicplinary and interdisciplinary forms of reasoning in astrobiological research as well.

There is a fluidity to these three sorts of cross-disciplinary engagements. Recognizing this fluidity inoculates us against assuming any single, strict typology might sufficiently characterize all the ways our cross-disciplinary engagements can proceed. Still, if we carefully pay attention to the shifting natures of our cross-disciplinary discourses, we can better generate substantial impacts at the boundaries of traditional disciplinary divides.[22]

This work represents a transdisciplinary theological project. It is committed to fostering mutual understanding that stretches transversally across disciplinary boundaries by thinking through how tenets of astrobiology intersect with various reflections on human ways of being in the world and belonging to the world. If identifying specific methods is helpful, consider my reflection in light of Celia Deane-Drummond's use of deep history,[23] Wentzel van Huyssteen's understanding of transversal reasoning,[24] and how astrobiology extends insights generated from the critiques of anthropocentrism in animal studies and theology—as with Deane-Drummond, David Clough, or Andrew Linzey.[25] However, a critical reader will not find my work strictly following any of the methods these thinkers develop nor offering a specific parallel to their previous arguments. Though clearly indebted to their works, I take this body of theological reflection as a ground for launching insights, particularly regarding the *imago Dei*, in a new direction.

In any case, the structure of the book is broadly inductive. The chapters provide a series of specific examples drawn from astrobiology, doctrinal reflection on the *imago Dei*, and reflections on the Anthropocene, to suggest an alternative approach to framing how human beings meaningfully are in the world and belong to it. Braiding together these diverse traditions, I suggest the Earth is not only a living planet but an artful one. To be an artful planet requires we take seriously geological history and the significance of the geological agency of *homo sapiens*. It also requires that we, as members of a species, own our responsibility for inducing new technobiogeochemical cycles into our planetary history.

*"Artful planet" is a transdisciplinary, aspirational symbol for how we live-into distinctively being an Anthropocene* imago Dei *in light*

*of our burgeoning astrobiological awareness.* It frames the meaningful relationship of self and world in light of cross-disciplinary constructs. As a theologian, though, I have a special concern for the role that theology—and particularly doctrine—can play in processes of transdisciplinary study that seek to bolster planetary ways of thinking. By creatively reclaiming doctrinal symbols, we can generate ontological language and ethical values that respond to the overwhelming sense of meaninglessness and anxiety generated by sensing that our individual agency is insufficient for dealing with the planetary crises we face. I call this a transreligious public theology of doctrine. I examine a particular doctrine or symbol from a specific tradition and historical location in order to discern how it distinctively frames a sense of ultimacy that may then inspire a public pursuit of transreligious values for a more just social order that reaches beyond the symbol's originating community to invite others to participate in ultimacy.[26] Doctrinal symbols, like the *imago Dei,* that have a wide cache to speak beyond a single religious tradition are particularly important for this process. As I will construe the *imago Dei* here, its actualization relies upon the participation of wider publics that go beyond any single faith community; and, it invites reflection on how doctrine can play a crucial role in thinking about the development of a more just social order.

Toward this end, Chapter 1 examines example astrobiological phenomena in order to move beyond popular preconceptions about astrobiology: that it is about studying aliens. Chapter 2 distills key themes from these examples (particularly "intra-action" and understanding life at a scale beyond the individual organism) to suggest how astrobiology can serve as a context of engagement for interpreting two symbols that are the focus of the rest of the book: the *imago Dei* and the Anthropocene.

Chapter 3 considers how the cosmogonies shaping the *imago Dei* intersect well with astrobiology's concern for intra-action. After examining inescapable themes from selected historical treatments of the image of God in Chapter 4, I offer a critical shift in the optic metaphor we use to think about "image": a shift from reflection to refraction. Paralleling this work, Chapters 5 and 6 explore how we understand the Anthropocene as a contemporary symbol for the intra-action of human being and nature that is corroborative to concerns related to intra-action in astrobiology and the refractive account of the *imago Dei.* Chapter 5 details how conceptualizations of "nature," particularly in an American context, that explicitly identify environmental degradation have been variously ignored. Chapter 6 then offers a constructive account of the meaning of the Anthropocene in light of the preceding work on astrobiology and the *imago Dei*: understanding ourselves as planetary creatures situated in deep time.

Finally, Chapter 7 weaves the themes of intra-action, refraction, planetarity, and deep time together. I propose that to be the *imago Dei* is not a property of individuals or even a species, but describes a categorical shift in planetary flows of energy and matter: a shift from biogeochemical cycles to *techno*biogeochemical cycles. If human beings live into our shared responsibility for being the *imago Dei*, then the Earth is not only a living planet, but an artful one.

Chapter 8 tackles a remaining, thorny problem. How can we adopt the account of the *imago Dei* put forward here without crushing the existential dignity of the individual? As individuals, we can only enact an existential "mood" or "attitude" for being-with creation that encourages a *possible* refraction of the divine and shapes the ways we, humanity understood as a particular type of living-system, come to be-with the world. To address this problem, I invoke three widely used themes in phenomenology—presence, wonder, and play—to describe what individuals pursue when the *imago Dei* is a planetary phenomenon. Finally, the brief epilogue attempts to give practical examples of how these themes of presence, wonder, and play might help us live into our shared responsibility as the *imago Dei* for meaningfully ordering our existence toward new possibilities of flourishing across creation—particularly with concern for those tiny aliens with which we may already be sharing the cosmos.

# CHAPTER

# 1

# Exoplanets and Icy Moons and Mars, Oh My!

What exactly is astrobiology? Pointing to a date when the study of astrobiology begins, as a specific field or form of inquiry, can be a surprisingly difficult task. The term itself has a sort of lineage. While it is broader than earlier terms used, such as xenobiology (biology based on foreign chemistry) or exobiology (a search for life and extra-terrestrial environments), it is certainly not unrelated to them. If the aim is simply to draw a line in the sand or note a seminal moment at which astrobiology research begins we might—at least loosely—point to NASA funding for exobiology research in 1959 or the establishment of the Exobiology Program in 1960 with its popularization by Joshua Lederberg. Yet the contemporary sense of astrobiology can hardly be understood without research in the 1980s and 1990s on extremophiles (organisms living in conditions originally thought to be uninhabitable). Perhaps, if we want a precise moment, we can point to the case made by Carl Woese, Otto Kandler, and Mark Wheelis in 1990 to rethink the taxonomic structure of biology in terms of three domains (Archaea, Bacteria, and Eucarya) as a beginning point: where a fundamental re-thinking in biology opens the door to the founding of astrobiology.[1] But, *astro*biology needs to get us beyond the earth, does it not? What if the seminal moment for this new field belongs to the discovery of the first confirmed exoplanets in 1992, around Pulsar B1257+12, and the subsequent onslaught of newly discovered planetary bodies in other solar systems.[2]

Because astrobiology is so diverse, it is difficult (if not impossible) to tell its history in terms of typical principles—such as the development of an object and method of study—as we might with more traditionally unified scientific fields.[3] Astrobiology brings together many distinct fields of scientific study but contextualizes them in terms of one another to address a particular problem and thereby begin shaping a style of inquiry not completely at home in any one of the participating fields. Thus, like so many other cross-disciplinary endeavors, it is challenging to try and categorize how exactly the various fields of study that are brought together under the heading of "astrobiology" actually intersect or whether all "astrobiologists" even agree about what this intersection might entail. A planetary scientist may consider astrobiology in very different ways from an evolutionary biologist, astrophysicist, or geologist who also participates in this cross-disciplinary inquiry.

If nothing else, I hope to dispel one popular misconception about astrobiology in this chapter: the commonsense or popular assumption that astrobiology deals specifically with aliens. For instance, when I started working on this project, a friend and chemistry professor quipped, "Great, you're going to study the 'science' where they are still looking for the object they want to study." Of course, as a theologian, I was unfazed because I cannot demonstrate the object of my own field of study exists! More to the point, the widely held supposition my friend invoked is that astrobiology is a study of alien life, but no such non-terran life is known to exist.

This widely held supposition needs to be challenged. Even simply acknowledging how NASA's astrobiology research is guided by three overarching questions (How does life begin and evolve? Does life exist elsewhere in the Universe? What is the future of life on Earth and beyond?) suggests that more than demonstrating and analyzing the existence of alien life is at stake.[4]

Here we will look at four examples of astrobiological phenomena: Kepler-452b, Proxima Centauri b, Enceladus, and Mars. Other examples could be chosen, but these four are particularly helpful in that they represent three distinct types of phenomena that are of astrobiological interest. Kepler-452b and Proxima Centauri b are exoplanets. Enceladus is an icy moon within our own solar system. Finally, Mars is our planetary neighbor and was previously quite possibly habitable.

My contention is that these examples are of *real astrobiological significance;* they are not simply placeholders for alien life scientists have yet to find. What emerges through a consideration of these astrobiological phenomena themselves is a vision of the cosmos committed to understanding the intra-action of living-systems and habitability. Working through these different sorts of examples, an astute reader will notice some common themes emerging: Astrobiology is very concerned with *microbial* life; it is concerned with living-systems more than individual organisms; and it is seeking clarity regarding how habitable conditions affect the occurrence and development of living-systems. I will analyze the social significance of these phenomena more explicitly in the next chapter, paying special attention to how the public presentation of findings in astrobiology might give us pause even if it *also* stresses that this research is about more than finding aliens. For now, my more modest hope is to expose, critically, the *interdisciplinary* factors that affect the scope and scale of this body of research, and make clear just how different this research is from astrobiology as it is popularly conceived.

## Kepler-452b

If there is one exoplanet that the wider public has heard about, it is probably Kepler-452b. Astronomers discovered it as part of the Kepler mission in 2015. Kepler is a space observatory and has been, since 2009, a primary telescope involved in discovering exoplanets by the transit method. It uses this method because Kepler's only instrument is a photometer: a device that measures light intensity. This photometer continually measures 145,000 main sequence stars in the constellation Cygnus.

The transit method, like all of our current means of finding exoplanets, is an indirect observation technique. We do not actually see the planet; we detect signs that indicate the given planet is almost certainly there. In the case of the transit method, Kepler is monitoring dips in starlight that result from an exoplanet passing between the observed star and the telescope. Once an exoplanet is discovered by the transit method, by continuously monitoring the dimming of starlight we can determine the orbital period of the exoplanet, estimate its distance from the star it orbits, and approximate its diameter in relation to the size of this star.[5]

It is worth bearing in mind that the dips in stellar output described here are *very small*. To detect an Earth-size exoplanet, Kepler has to register a 0.01 percent dip in the intensity of starlight for at least three transits (less than three dips could indicate a false positive); *then* the finding must be confirmed using an additional indirect technique for discovering exoplanets or subsequent ground observations. Moreover, observing such a dip at all relies on an exoplanetary system orbiting its star in, virtually, the same plane as our own solar system. (By contrast, imagine a system where the exoplanets orbit their star on a plane offset 90 degrees to our own; there would be no dip in starlight because the exoplanet would not pass between its star and us.)[6]

At the time of its discovery, Kepler-452b was something novel: It was only the sixth "super-Earth" (a planet with a radius less than two times that of the Earth) found within the conservative habitable zone of its star.[7] Moreover, it had been in the habitable zone of its star for 6 billion years. In short, it had been in an orbit where the presence of surface water and an atmosphere could not be ruled out for a longer time than the Earth has had life on it.[8]

In addition, it boasted a striking number of similarities to the Earth. Kepler-452b orbits a G2-type star, one like our sun, every 385 days. This means it is only 5 percent farther from its star than Earth is from the Sun. Still, this "Earth cousin," as popular reporting frequently dubbed it, is likely to be a little warmer than our planet because its star, Kepler-452, is 20 percent brighter and 10 percent larger in diameter than our Sun. This similarity of Kepler-452 to our own Sun proved important for the initial announcement of the discovery because it allowed for speculation about similarities between Earth and this exoplanet that made for headline grabbing sensationalism. As Dr. Daniel Brown at Nottingham Trent University was quoted, "This is so fascinating because Kepler-452b receives the same kind of spectrum and intensity of light as we do on Earth. This means plants from our planet could grow there if it were rocky and had an atmosphere. You could even get a healthy tan like here on holiday."[9]

After the initial flurry of reporting on Kepler-452b, enthusiasm for the planet cooled considerably. Because it was discovered by the transit method, crucial information about Kepler-452b's mass (information that would be required to confirm it was actually rocky and not just a small gas giant) could not be obtained. At the time, there was also no way to do follow-up observations via astrometry or radial velocity measurements to determine the mass of the planet. However, the initial article announcing the discovery suggested that there was still a 49–62 percent likelihood that the planet had a rocky composition given probabilistic forecasting relating planetary radii and masses. Subsequent forecasting decreased this probability significantly, to 13 percent, suggesting Kepler-452b might be much more like Neptune than Earth.[10]

Additionally, public excitement also began to wane because the star this exoplanet orbits is approximately 1,400 light years away. To put that in context, humans would need to have shot *a probe moving at the speed of light* into space as part of the first Olympiad in 776 BCE for us to be expecting its return within the next decade. Alternatively, with the speed of current unmanned probes, it would take approximately 26 million years to reach this older Earth cousin.

We can reasonably suspect that we will know more about Kepler-452b soon. The Characterizing Exoplanets Satellite (CHEOPS) mission will gather information about the size of exoplanets for which we already know the mass, increasing the number of data points used in forecasting models. This will allow better predictions about whether planets observed by direct transit are likely to be rocky. Moreover, Kepler-452b remains a good candidate for observation by the James Webb Space Telescope in order to identify its likely atmospheric gases. Even if it turns out that Kepler-452b is not so much an Earth cousin but something more like a third cousin twice removed (perhaps a sub-Neptune with a primarily hydrogen or helium atmosphere instead of a super-Earth), it will likely remain an interesting object of study—especially for planetary scientists interested in issues of formation. It is interesting because despite the many similarities to Earth that made it so tantalizing in the first place, it seems as though Kepler-452b might be far more different from our planet than was first assumed.

It is so different that recent analysis suggests it might not even exist! Arguing against confirmation of exoplanets by statistical means and the possibility of signal-to-noise disruptions, a new paper in *The Astrophysical Journal* contends Kepler-452b may have only a 16 percent chance of existing if the signal-to-noise threshold used to analyze Kepler mission data is changed. Kepler-452b is particularly vulnerable to such a shift because its period of orbit is so long (much like our own) compared to other exoplanets that we have discovered with an orbital period of less than two hundred days.[11] Whether "Earth cousin" or nonexistent statistical blip, Kepler-452b has been and remains a significant object of concern for astrobiology.

## Proxima Centauri b

Next, consider Proxima Centauri b. It is the closest Earth-sized exoplanet to our solar system found so far. Discovered in 2016, it was found using a different detection method than that used for Kepler-452b: radial velocity. This method, sometimes called Doppler spectroscopy, was the first used to detect exoplanets and remains a critical tool for confirming the existence of these other worlds. It relies on the fact that stars do not remain perfectly still when they are orbited by a planet. The star moves very slightly in a circle or an ellipse around its center of mass because of the gravitational pull of the orbiting planet. It is as though the star wobbles in the sky. In our own solar system, the Sun evidences just such a wobbling that coincides with Jupiter's twelve-year orbit and Saturn's twenty-nine-year orbit.

These small shifts have an effect on a star's spectral lines. These lines, recorded by a spectrograph, are like a barcode for a star. The spectrograph refracts white light from a given star into a frequency spectrum (imagine a band of rainbow colors) and records it. If there were nothing between you and the star, you would see a continuous spectrum (an uninterrupted band). However, this is not what we see.

Instead, a frequency spectrum with dark bands in it appears. These dark bands represent the presence of specific elements in the stellar atmosphere of the star. In the case of Proxima Centauri (the star around which we can find Proxima Centauri b) there is a strong spectral line at a wavelength of 280nm indicative of ionized magnesium.[12] Each star will have dark bands in its spectrum that correspond to the elements present in its stellar atmosphere, giving us a distinctive barcode or fingerprint for the star.

If a star has an orbiting planet and we compare the spectral lines over time, the position of the dark bands in the spectrum will shift slightly. This is not because the stellar atmosphere has suddenly changed. It is a result of the Doppler Effect: an emitted frequency becoming higher as it moves toward an observer or lower as it moves away from an observer. (This same effect accounts for the change in pitch we perceive as emergency vehicle sirens pass by us.) Because the star with an orbiting planet is moving very slightly around its center of mass, the spectral lines of the star will move higher (or bluer) in the spectrum as the star is moving toward the Earth and lower (or redder) in the spectrum as the star is moving away from the Earth.

If we observe a repeating pattern of red-shifting and blue-shifting in the spectral lines of a star, this indicates the presence of a planet. How often the pattern repeats allows us to estimate the orbital period of the planet. Additionally, the size of the shift allows us to surmise a minimum mass for the orbiting planet (*i.e.* while the mass could be greater, the planet's mass cannot be less than the indicated amount given the degree of red-shifting and blue-shifting observed).[13]

Because Proxima Centauri b was detected by the radial velocity method, our knowledge of it is somewhat limited. Scientists estimate its minimum mass to be 1.3 times that of the Earth, and its orbital period is only 11.2 days. Thus, while Proxima Centauri b is likely to have a density similar to that of the Earth, it is far closer to its star than the Earth is to the Sun. Nonetheless, it falls within the conservative habitable zone for its star because Proxima Centauri is an M-type star (a red dwarf) that is approximately 2,700 degrees Kelvin cooler than the Sun (a G-type star) and only 0.1 percent as bright.[14]

Unfortunately, subsequent observation has indicated astronomers cannot also detect Proxima Centauri b by the transit method. This would have allowed scientists to determine the size of Proxima Centauri b and thereby make a better inference about its density.[15] Instead, working only from the minimum mass indicates that while Proxima Centauri b may be Earth-sized, it would be hard to say it is Earth-like.

While it seems likely that Proxima Centauri b is rocky,[16] the presence of water on the planet is difficult to predict. Though being in the habitable zone indicates that it is possible for liquid water to be on the surface of the planet, the likelihood that any water is present at all would depend on its planetary formation. If Proxima Centauri b formed where it is, it is unlikely that it has any water on it; a planet so close to its star would heat the water to vapor during its formation, and it would not be retained on the planet.[17] However, if the planet formed farther out while the gas and dust disk birthing the planetary system remained and then migrated into its current orbital location, it is more likely to possess substantial water because it would have initially gathered water in the form of ice.

Moreover, one of the great challenges to forming life on planets orbiting M class, red dwarf stars are flares. A flare is a brief eruption in the surface of the star that emits protons that would quite possibly destroy any atmospheric conditions that would protect life from harmful, high-energy UV light.[18] Proxima Centauri flares more frequently than our sun and Proxima Centauri b is far closer to its star; the exoplanet has likely been repeatedly and severely exposed to atmosphere destroying flares.[19] If Proxima Centauri b had a strong magnetic field, it could deflect these flares and potentially have an atmosphere. Such a possibility relies on the supposition, though, that a tidally locked planet (a planet that keeps the same face perpetually directed toward its star—so one side is permanently in the day and the other is permanently in the night) can produce such a magnetic field by convection in the planetary core produced from tidal heating by the gravitational pull on the planet.[20]

Despite the uncertainties of what Proxima Centauri b may be like, news coverage of the announcement of this discovery stressed not only its proximity to Earth, but also the importance of it being in orbit around an M-type star. Not just any M-type star, though. Proxima Centauri is the nearest star to Earth besides the Sun. Located 4.22 light years away, exoplanets found around Proxima Centauri are the closest we will ever discover.[21] While with the current speeds of unmanned probes it would still take approximately 54,000 years to reach Proxima Centauri b, public announcements about the exoplanet often also referenced the Breakthrough Starshot Initiative.[22]

Backed with 100 million dollars and initial support by celebrity scientists such as Stephen Hawking, this independent initiative is mostly following a conceptual approach laid out by Philip Lubin.[23] The aim is to propel very small probes (approximately 1 centimeter in size) equipped with thin sails via a powerful (100 gigawatt) Earth-based laser, thereby accelerating these nanoprobes to approximately 20 percent of the speed of light. The principle behind the idea is that a solar sail will allow for interstellar travel without relying on traditional rocket propulsion—an idea already in principle employed by JAXA's IKAROS spacecraft.[24] However, the Breakthrough project is imagined to have a far smaller sail (only 4 meters wide) and the laser would provide far more power than the sunlight propelling IKAROS.

The Breakthrough Starshot Initiative is still science fiction; it is a promising idea, but there are significant technical challenges that remain unaddressed. Notably, a 100-gigawatt laser would be 1 million times more powerful than today's continuous lasers. Even with such a laser, other technical challenges abound. If the Starshot sail did not remain perfectly flat while being bombarded by the laser, the craft would fly drastically off course; how does one keep the sail perfectly flat? How could we prevent a collision with particles in the interstellar medium or the Oort Cloud as the probe leaves our solar system, which would be catastrophic for the craft? Moreover, even if the Starshot craft reached Proxima Centauri b, it would not suddenly slow down when it got there.

This final point is particularly vexing. The craft would only be in the Proxima Centauri system for around two hours getting within, approximately, one astronomical unit (the distance from the Earth to the Sun) of Proxima Centauri b. Practically, this means that to get a flyby image of Proxima Centauri b, the one-centimeter chip on the sail would need a camera (not to mention whatever other instrumentation we might want to squeeze onto this tiny surface) capable of taking a relatively clear image while moving at 60,000 kilometers per second and then manage to transmit that information back to Earth.

Despite all the technical challenges to be overcome, if the Starshot system were completed, it would only take a probe approximately 25 years to reach Proxima Centauri b and then an additional 4.22 years to send information back to Earth. While this may seem like a significant investment in time, this is *drastically* quicker than reaching Proxima Centauri b with conventional rocketry. It is tantalizing to imagine the possibility of getting detailed flyby information—perhaps even a surface view—over the course of a generation (as with Starshot) rather than over the course of human history from the Upper Paleolithic to the present (as with conventional rocketry).

News coverage also significantly concerned itself with the type of star Proxima Centauri b orbits because more than 70 percent of the stars in our galaxy are also M-type stars. These stars are significantly cooler and dimmer than G-type stars like the Sun. If life were sustainable on a planet like Proxima Centauri b, this would be highly significant as regards the potential findings of NASA's Transiting Exoplanet Survey Satellite (TESS) mission launched in 2018. The goal of TESS is to find Earth-sized planets in the habitable zone of nearby stars. The mission timeline of two years means TESS will not be searching for Sun-like stars; but, if life is sustainable on planets around M-type stars, then the TESS mission could provide many new targets for searching out non-terran life. The recent finding of seven Earth-like planets (including three in the habitable zone) in the Trappist-1 system (an ultracool M-type star forty light years away) and further statistical analysis with revised planet-occurrence models continue to indicate that this area of astrobiological research is highly compelling.[25]

Additionally, even if it is found that planets like Proxima Centauri b orbiting M-type stars do not often have life on them because of how they were formed but they are still able to sustain life, such planets could eventually become prime candidates for colonization because M-type stars burn far longer than G-type stars. While the Sun will burn out in around 4 billion years, Proxima Centauri is likely to burn for an additional 4 *trillion* years.[26] If habitable planets exist around M-type stars, these could be likely targets for interstellar travel if that ever became possible. It does not take too many additional findings like those of Proxima Centauri b, Ross-128b, or the Trappist-1 system for our science-fiction imaginations to kick into overdrive: envisioning an interstellar highway system traversed by non-rocket technology with exoplanet rest stops at the various M-type stars along the way.[27]

### Enceladus

Astrobiology is not only concerned with exoplanets; it also is examining many objects in our own solar system. Icy moons in the outer solar system, such as Europa, Titan, and Enceladus, have been of interest to astrobiology for many years. Yet icy moons hold a different sort of astrobiological interest than exoplanets because we can study and observe these objects of our own solar system in detail. Enceladus provides a ready example of this difference.

Enceladus has been making news recently. It is one of Saturn's sixty-two moons and the sixth largest. While in the top 10 percent for size amidst Saturn's moons, this is a deceptive description; with an approximately 310-mile diameter, its total surface area would fit inside five U.S. states (Kansas, Nebraska, Iowa, Missouri, and Arkansas). Anyone might reasonably expect Enceladus simply to be an icy, small planetoid that is quite unremarkable; given its size, it should have a cold interior with minimal or no geological activity. It should hardly be newsworthy.

It has proven, nonetheless, surprising. Enceladus is a dynamic planetoid with significant thermal activity in the vicinity of its southern pole. While impact craters and some tectonic features have been noted to be present on Enceladus since the Voyager 2 fly-by in the early 1980s, more recent data has increased interest in the tectonic features of the moon.[28] At its southern pole, there are parallel fractures—sometimes called "tiger stripes"—where plumes of gas and misty ice vent material from the moon into space, at times to a distance three times that of the diameter of the moon.

These plumes, detected in 2005, are what have truly made Enceladus notable: so notable that NASA made changes to the Cassini spacecraft's mission to explore Saturn (which initially discovered the plumes) so that it would pass by Enceladus twenty-two times. Beyond the very existence of the plumes, which have also been found on Europa, much of the excitement arose over what was in them: They suggest that the interior of Enceladus may provide three crucial requirements for life—liquid water, an energy gradient, and chemistry that could plausibly be linked to biotic entities.[29]

The plumes are thought to be produced by a liquid water ocean underneath the icy surface of Enceladus. The ocean has been posited because of the detection of salts in ice grains gathered during Cassini's plume samplings. These salts are important. Research indicates the outer shell of Enceladus is pure water ice. Thus, if salts appear in the ice grains sampled from the plume, then the outer shell of the moon cannot be the sole source of water feeding the plumes.[30]

The size of this sub-ice ocean remains a bit of a mystery. At the least, scientists suggest we should hypothesize the presence of an internal ocean across the southern hemisphere of the moon, even if it does not cover the entire global subsurface. An ocean across the southern hemisphere of the moon would be consistent with tidally induced heat dissipation, which helps to explain significant differences in the observed surface temperature of the moon (ranging from −340 degrees Fahrenheit at the equatorial surface to a range of −100 to −130 degrees Fahrenheit at the south pole).[31]

However, recent theories suggest that this sub-ice ocean may not be as localized, limited to one hemisphere, as was initially thought. Additional research suggests the moon has a layered internal structure with a rocky core, liquid ocean, and icy surface. Scientists estimate that the rocky core accounts for approximately 65 percent of the moon's material (more than previously assumed) while the liquid ocean and ice compose the other 35 percent. This global subsurface ocean is thought to be the product of tectonic forces resulting from Enceladus's gravity field and libration due to its gravitational interactions with at least one other moon, Dione.[32] The varying gravitational force exerted by Saturn and its other moons on Enceladus through the course of its elliptical orbit would not only account for a global sub-surface ocean, but it has also been suggested as a means to accounting for significant changes in the amount of material that spews forth from the plumes at different points in Enceladus's orbit around Saturn.[33] This flexing of the core of the moon through its orbit and the presence of a global sub-surface ocean would provide significant liquid water and a persistent source of energy for the development of a biotic system.

Moreover, Enceladus's plumes have proven to be particularly intriguing in terms of not only explaining why they exist, but also understanding how the chemical compounds found in them came to be there. Gaseous compounds found in the fly-bys of Cassini suggest that critical elements used by all life on Earth (the so-called CHNOPS elements—carbon, hydrogen, nitrogen, oxygen, phosphorous, and sulfur) would be in the ocean or core of this moon.[34] While none of the compounds discovered so far represent a distinct biosignature, many of those present, including carbon dioxide, water, methane, molecular nitrogen, propane, acetylene, formaldehyde, and ammonia, are associated with life. Some have even suggested that longer chain carbon compounds are present in the plume but have broken into smaller compounds (consistent with the behavior of longer chain organics) upon impact with Cassini's instrumentation as it sampled the plume.[35]

Given the liquid water, tectonic activity, and compounds discovered in the plume, one could even readily imagine the sort of life that Enceladus might support, as it would have a clear analogue to specific extremophiles here on Earth: the microorganism communities found at Lost City. The comparison *specifically* to Lost City is intentional because, as a hydrothermal system, it is quite different from the "black smokers" that more typically come to mind.

The difference is simple to illustrate. Black smokers form at areas where tectonic plates are drifting apart. As they drift apart, superheated material from the vent fluid containing various minerals (especially sulfides) meets near freezing seawater. The minerals then precipitate into column-like structures. Black smokers are highly acidic hydrothermal systems (having a pH of approximately 3). By contrast, Lost City is an alkaline hydrothermal system (with a pH of approximately 9–11) that works by a different process: serpentinization. Serpentinization is a type of metamorphic, exothermic reaction. It takes place at low temperatures, geologically speaking, and oxidizes and hydrolyzes ultramafic rock. These different processes release different gases (hydrogen and methane at Lost City in comparison to primarily hydrogen sulfide at black smokers) and provide different energy gradients for the chemoautotrophic forms of life in these habitats. Given the significant differences in pH and released gases, scientists find distinct sorts of chemoautotrophic microbial communities in these alternative hydrothermal systems (the chemoautotrophs at Lost City would not survive at a black smoker, and vice versa). Perhaps unsurprisingly, these two hydrothermal systems support very different sorts of living things; for instance, there is a lack of larger animals (like the tube worms of black smokers) and a preponderance of biofilms inside the chondritic rock vents at Lost City.

The astrobiological significance of distinguishing these hydrothermal systems again relates to the analysis of Enceladus's plumes. The plumes are highly alkaline (with a pH of approximately 11 or 12) with significant volumes of dissolved sodium carbonate. This would be consistent with the ongoing serpentinization of chondritic rock in the sub-ice ocean of Enceladus. Now the similarities are becoming striking and doing predicative modeling about the subsurface ocean of Enceladus based on analogy to chemical oceanography here on Earth, specifically drawing an analogy to the Lost City hydrothermal system, seems quite reasonable. Recent findings of molecular hydrogen in ratio to carbon dioxide and methane that are indicative of a thermodynamic disequilibrium that would be well explained by the presence of hydrothermal vents on Enceladus significantly buoys such an effort.[36]

Just as we find here, we would expect serpentinization on Enceladus to produce significant energy gradients that could support chemoautotrophic carbonate life-forms. Subsisting on hydrogen and methane, such hypothetical life-forms would not be reliant on photosynthetic energy cycles.[37] Moreover, some scholars working on the origins of life on Earth have proposed that the abiotic geochemistry of a process like serpentinization also provides a certain level of physical compartmentalization that would be necessary for the abiogenesis of some form of chemoautotrophic life.[38] In short, Enceladus seems to have the right sort of chemistry available to make a parallel to one means of imagining the formation of simple life on Earth.

In principle, as popular reporting on Enceladus has indicated, while discovering life was not part of the Cassini mission agenda, given additional theorization regarding the redox chemistry of this moon and the recent discovery of a few additional elements in the plumes (particularly molecular hydrogen), the stage may be set for a return mission that would look to detect samples of life in the geyser emissions instead of having to imagine how one might drill through surface ice to take samples.[39] This in-principle argument may be even more significant for the upcoming Europa Clipper mission that is slated to launch in the 2020s. More recent analysis of Galileo mission data taken in the late 1990s seems to confirm a suggestion first made in 2016 that plumes are also erupting from Europa, a moon of Jupiter.[40] There is less data about the content of these plumes, but the same principle of sampling geyser emissions instead of drilling through ice to take samples may be able to be used in studying this other icy-moon.

## Mars

If there is one site of astrobiological significance that has captured the public imagination almost more than any other, it must be Mars. Part of the reason for this is that Mars falls just within the habitable zone of the Sun by some calculations, and significant portions of its surface were likely covered in liquid water in Mars's past. All this makes Mars a distinctive astrobiological phenomenon because even if it does not have life on it at present, it may have had life on it *in the past*.

Generally, we divide Mars's geological history into three eras: Noachian, Hesperian, and Amazonian. Giving definitive transition points by which Mars moved from one era to another is exceptionally difficult.[41] Nonetheless, the three eras provide a helpful guideline for thinking about Mars's habitability. From the Noachian to the Amazonian, there is a steady decrease in the amount and types of available surface water. In the Noachian, the oldest era (4.5–3.7 Ga), there are several lines of evidence of fluvial (rivers and streams) and potential pluvial (rainfall) activity in the geological record, including the presence of clay minerals, carbonates, and aqueous minerals. Whatever fluvial activity may have been occurring, these lines of evidence indicate the surface water present was potentially habitable and likely to be of a neutral to alkaline pH balance.[42]

However, the likelihood and extent of pluvial activity remains debated, in large part because it is difficult to tell just how wet and warm the climate of Mars during this era would have been. In particular, we need to better understand how the atmospheric conditions during this time relate to the density of observed valley networks. In short, could precipitation similar to that found on arid regions of the

Earth have produced these valley networks as opposed to more epi-sodic periods of melt off from ice sheets?[43] While there is evidence of moving water in the geological record, there is no guarantee that the fluvial activity lasted for an extended, continuous period. However, being clear about the climate conditions is important because it not only affects interpretation of the geological record, but it also influ-ences what we might assume to be the biotic potential of Mars dur-ing this older age. Understanding Martian climate conditions during this era would give scientists a better sense for how extensive the habitable conditions for life may have been.

In the subsequent Hesperian (3.7–3.0 Ga) and Amazonian (3.0 Ga to the present) eras, signs of water involve significantly less volume and become increasingly sparse. Nonetheless, during the Hesperian era, wet conditions may have continued to exist regionally and in-termittently largely due to outbursts from subsurface aquifers. At the least, there is continued evidence of aqueous alteration of min-erals through the Hesperian era.[44]

The Amazonian period, however, is even more dry than the Hes-perian. Despite this trajectory toward dryness, some evidence of liq-uid water, often a brine, persists even today. Some researchers have proposed that water layers a few nanometers in depth could exist between mineral grains. More pronounced flow-like features (called recurring slope lineae—RSL) have recently been cited as evidence of ongoing hydrological activity on Mars that ebbs and flows with warmer and cooler seasons at four separate sites. The additional presence of hydrated salts when these flows are at their widest and the suggested potential for an exchange of water at the atmosphere-soil interface at a different site all suggest a surprising degree of on-going water activity.[45]

What should immediately stand out about Mars as an astrobio-logical phenomenon is that, in looking for the presence of liquid wa-ter, we have to take into account its geological history. It is not that we discount this feature with regard to icy moons or exoplanets, but in the case of our solar system neighbor, we simply have signifi-cantly more knowledge about its planetary history. Thus, when we examine Mars we are looking for its biotic potential *both* during the Noachian era *and* in the present. In the case of Mars, we are search-ing for life that may have been and that still might be.

The context of those two searches are very different, however. In searching for life that may have been, we are undertaking a sort of bio-archaeology project. As with the study of life forms long dead on this planet, the search for indicators of life that would have been present on early Mars relies on fossilized evidence of life or indisputable geological markers of life interpreted in light of what we can surmise about Mars's significant climatic changes.

Imagine going back in time, though, and searching actively for life on Mars during the Noachian era. What would you be looking for? The question is not as frivolous as it might seem at first; being able to answer this question guides what sort of fossilized or geological evidence we seek today.

During the Noachian era, there were large-scale bodies of water and potential climates sufficiently habitable to various types of extremophiles. Based on a model of abiogenesis that could be parallel to the early Earth or to the life cycle of various extremophiles still living on Earth today, a search for life on early Mars would be, in principle, similar to our current search for life on icy moons in the outer solar system. In both cases, the hope is to find life arising in large bodies of water with sufficient energy resources and an appropriate chemistry.[46]

It is in the context of the search for previous life on Mars that one of the most infamous astrobiological phenomena came to prominence: the meteorite ALH84001. The meteorite was a Noachian era igneous rock that flew from Mars to Earth because of impacts on Mars. Essentially, about 4.1 billion years ago, a large object (like an asteroid) hit Mars hard enough that pieces of rock from the planet flew off the surface and then hit the Earth. ALH 84001 made headlines in 1996 because of the presence of alleged microfossils in the rock along with other mineral and chemical indicators (carbonates, polycyclic aromatic hydrocarbons, and magnetite) suggesting there was previously Martian microbial life.[47] The headlines were sufficiently motivating that then President Bill Clinton made a formal set of comments with the televised announcement of the finding.

Eventually, nearly all of the scientific community determined ALH84001 did not provide evidence of Martian biota (that is, no tiny aliens). Abiotic processes could explain the presence of all the chemical and mineral indicators. Moreover, the alleged microfossils were by no means definitive; some inorganic mineral surfaces can produce rod-like structures that appear like microbe fossils. In this case, the rod-like structures are around ten times *smaller* than similar terrestrial microbes, thereby bolstering the appeal of an abiotic explanation.[48] The meteorite remains a testament to just how difficult it is to incontrovertibly demonstrate microbiological activity from a fossilized feature. A certain degree of skepticism is warranted in the face of any claim to find evidence for microbial life in a bygone geological era, even if the process we imagine could have taken place to give rise to life seems very familiar.

The search for current life on Mars has had to take a very different turn from these searches for evidence of life in Mars's past. This is because we do not find large bodies of liquid water and there is not a sufficient atmosphere to shield life on the surface of Mars from solar radiation. In addition, the low temperatures and low pressures would inhibit metabolic repair for a microorganism on the surface; yet active metabolic repair would likely be necessary for any organism so pervasively exposed to solar radiation.[49] Simply put, the conditions one might need for the emergence of various sorts of chemoautotrophs (the sorts of life sought out on Enceladus or fossilized as part of Mars's early planetary history) are just not present on Mars today.

However, this does not mean that Mars's current astrobiological significance is simply mute. Even a mere one-meter rock layer would be sufficient to block high-energy particles and a millimeter-thick rock layer can provide shielding from UV radiation. With such shielding, some sort of life analogous to chemoautotrophs could exist, conceivably, under the surface of the planet. The crucial issue, then, would be to ascertain how small or ephemeral an aqueous environment can be and still provide a sufficient solvent and energy gradient for a biotic system to exist. That is still a question of some debate.[50] If we remain optimistic, a search for current life on Mars would need to look for subsurface, *small-body* aqueous environments with some sort of redox gradient that could support lifeforms like various chemoautotrophs on Earth. Finding such life would likely serve as a proof of concept for just how little water is needed for life to arise.

Astrobiology is highly interdisciplinary and highly complicated. If these locations in the cosmos are of direct astrobiological interest, or if these are the sorts of phenomena that are astrobiologically significant, then what astrobiology is studying and seeking to demonstrate is something quite distant from commonsense, popular conceptions that link astrobiology directly to the study of alien life that has not yet been demonstrated to exist. Keeping this in mind, we need to move one step further: How do we characterize the significance of astrobiology once we question the commonsense approach? It is to this question we must now turn.

# CHAPTER

# 2

# Astrobiology's Intra-Active Aliens

This chapter will not solve the complex issue of how astrobiology's academic self-identity ought to be parceled out. Instead, I want to consider only how we frame the social significance of astrobiology when we move beyond popular preconception. Looking at astrobiological phenomena in the last chapter that are of *direct* interest to astrobiologists, we find the study of astrobiology is about something both much broader and simultaneously much narrower than a generic search for extraterrestrial life.

The study of astrobiology is narrower in that rigorous documents defining the scope of the field, such as the *Astrobiology Strategy* from NASA, evidence a commitment to searching out chemistries for life congruent to life *as we currently know it*. This helps restrict searches for what constitutes a relevant astrobiological phenomena to more definite targets.[1]

While it may be narrower in scope, we also find that the scale of consideration is much larger than in popular conceptions of astrobiology. By looking at example astrobiological phenomena in the previous chapter, we gained a sense of the *interdisciplinary* factors that affect the scope and scale of this body of research; issues of astrophysics, geology, and chemistry are inextricable from any adequate conceptualization of astrobiology. This field of study requires interdisciplinary engagements that go well beyond specializations we assume to be relevant. Instead, what emerges through the consideration of astrobiological phenomena themselves is a style of envisioning the cosmos *that is committed to understanding the intra-action of living-systems and habitability*. One can reasonably argue that a central concern of astrobiology is establishing an adequate general theory of living-systems at the level of populations, irrespective of whether we ever confirm or deny the existence of alien lifeforms.

Technically speaking, these interdisciplinary discoveries of astrobiology considered from the last chapter proffer a particular *transdisciplinary* vision of the cosmos. Further, this transdisciplinary vision has interesting theological relevance as a context for doing theological anthropology and intersects with a certain strain of environmental thinking about the Anthropocene. Examining how this specifically is the case will be the focus of the four subsequent chapters.

The goal of this chapter is to lay a conceptual foundation for the subsequent chapters by introducing how the intra-active quality of astrobiological phenomena provides an important context for framing what constitutes meaningful human existence because the scale of these phenomena is vast. To think astrobiologically requires that we imagine significant ontological units beyond the human individual and her agency that accord with the more general theory of living-systems that astrobiology is beginning to articulate.

## More Than Just Waiting for Aliens . . .

Whether describing exoplanets, icy moons, or our next door neighbor in the solar system, each of these examples, especially the public presentation of the scientific work entailed by each example, has something in common: the tendency of new discoveries in outer space to have a kind of flash-in-the-pan moment. We all think, "That's cool!" and then we click onto the next story in our ever-lengthening newsfeed. Whether it is about the discovery of the newest exoplanet or claims about microbes on Mars, there is a sense in which this pattern seems to hold. There is risk in this pattern.

On the one hand, it seems undeniable that there are advantages for this burgeoning scientific inquiry to be in the public eye. For instance, Kepler-452b remains the most familiar exoplanet to the wider public (or lingers in our communal pop-cultural memory) despite the fact that its moniker as "Earth's older cousin" was highly hyperbolic, and the exoplanet itself may not even exist. Likewise, and perhaps in even more spectacular fashion, the claims about Martian biota in ALH84001 were a tremendous boon to burgeoning astrobiology research programs. As Everett Gibson, who worked on the initial research regarding the meteorite, and Andrew Steele, an astrobiologist at the Carnegie Institution for Science, have each reflected, without this event, astrobiology might never have become so popular, and significant funding for Mars exploration programs would likely have not been reinstated.[2] Astrobiology must intersect with the public imagination to continue generating support for the significant research funding required to pursue such inquiry.

Nonetheless, if astrobiology continues to be presented this way to the wider public, it is at risk of developing a bad rap as simple sensationalism. Of course, this tendency toward sensationalism relies upon some implicit expectations about what astrobiology *is*. Given what has already been claimed about public expectations regarding astrobiology, it is no wonder that astrobiological phenomena like those considered in the previous chapter are destined to disappoint. The example astrobiological phenomena I described would only seem to be important *indirectly*: buoying the hope that we will find aliens eventually.[3]

Yet for the astrobiologist, these example phenomena are of *direct* interest; they are not merely placeholders for an undiscovered alien being. To understand why this is the case requires thinking about astrobiology beyond its immediate, popular sensibility. One way to press beyond the popular conception is to examine NASA's *Astrobiology Strategy*.[4] The *Strategy* names six major objectives of research that focus on what is at stake in connecting the "astro" and the "bio" aspects of this inquiry. I would suggest these six objectives reflect three broad themes relevant to the wider goals of this project: avoiding the abiotic, understanding living-systems, and conceptualizing habitability.

We might take this theme of "avoiding the abiotic" in two senses. First, it indicates that astrobiology must be sure to eliminate false positives—to ensure that *any* potential biosignature is not from an abiotic source. The scrutiny of ALH84001 is a good example of putting this theme to work. In this sense, the theme is methodologically significant as a broad, apophatic principle guiding what sort of objects are of concern to this cross-disciplinary inquiry: If something is an astrobiologically relevant phenomenon, one has to ensure that the relevant features are *biologically* and not just astronomically significant.

How exactly does one avoid the abiotic though? How do we assuredly draw a line between living and non-living, or between potentially significant for life and not possibly significant for life? After all, life may be utterly different from our own—perhaps using an alternative solvent like methane instead of water or a different set of essential elements than the ever-popular CHNOPS elements crucial to our own biotic chemistry. While all of this may be possible, we realistically still only have one example of life coming about in one spot in the cosmos. When your sample size is one, extrapolating what exactly you are looking for when you say you are searching for life—in any form it might take—elsewhere in the universe can be a bit dicey.

In this sense, the theme "avoiding the abiotic" has a second, narrower sense as well that the first objective from the *Strategy* document helps to clarify: "Identifying abiotic sources of organic compounds." The language from the *Strategy* document is specific; it appeals to "organic compounds." This is definitely intentional. I would suggest this reference to "organic compounds" represents a narrowing of the scope of the objectives of this cross-disciplinary inquiry to chemistry suitable for life *as we currently know it*. The focus of astrobiology is on finding life that is, in a most basic sense, analogous to the only known example we have for life arising somewhere in the universe—life here on Earth.[5]

This approach narrows our search through the vastness of space by giving significantly more definite targets for what we are seeking to find. In all the examples from the previous chapter, water is the assumed solvent or medium to be sought wherein biotic chemistry can occur. Subsequent to water, astrobiology searches for various organic compounds near these water-rich environments. This crucially delineates how a phenomenon is of astrobiological, not just astronomical, interest. When we tie the types of biosignatures, signs of life, we are looking for in the universe to the search for water and organic compounds, we can be much clearer about what it means to eliminate false positives. The aim is to search out planetary bodies with liquid water that have no known abiotic means to explain the disproportionate presence of organic compounds.

If the theme of "avoiding the abiotic" is largely methodological (concerned with how astrobiological inquiry ought to proceed) then the final two themes, "understanding living-systems" and "conceptualizing habitability," are more concerned with astrobiology's object of study. Carl Pilcher, the former director of the NASA Astrobiology Institute, has succinctly suggested an excellent one-line definition of astrobiology that puts these final two themes in intimate relation to one another. He observes that astrobiology is "a quest to understand the potential of the universe to harbor life beyond Earth."[6]

One of the most helpful aspects of Pilcher's definition is that it rules out the implicit public expectation that astrobiology is about studying alien life. His definition certainly makes clear that finding an alien would be excellent, and he gives some clear answers about how this astrobiological quest might proceed, but finding an alien lifeform is neither the ultimate nor the entire aim of astrobiology. Instead, it is concerned with a "potential" of the universe. Astrobiology is inquiring into the *possibilities* and *conditions* by which life might emerge.

The type of "life beyond Earth" being sought in Pilcher's definition can all too easily be misconstrued in the popular imagination. The aliens invoked by Pilcher are not the intelligent beings of science fiction narratives or the communicative aliens sought by projects such as SETI. Instead, invoking two of the objectives of the *Strategy* document hones what life beyond earth entails here: the "synthesis and function of macromolecules in the origin of life" and the study of "early life and increasing complexity." If astrobiology is dealing with macromolecules and early life before it fans out into seemingly preposterous complexity, then the interest is in *microbial* life: teeny tiny aliens. This concern for microbial life is borne out by each of the example phenomena—though the cases of Enceladus and Mars, where the microbial alien life sought is analogous to the metabolism of chemoautotrophs, exemplify this particularly well.

What is equally curious, however, is the *absence* of an objective that we might assume would appear in the *Strategy* document: to define life. On the one hand, this might be because we already find a working definition of life in the *Strategy* document itself: Life is "a self-sustaining chemical system capable of Darwinian evolution."[7] Of course, the focus on chemical systems in this definition further emphasizes just how exceptionally simple the life, or even proto-life, considered in astrobiology is. More importantly, however, invoking Darwinian evolution indicates that what astrobiologists seek is not an *individual* living thing but *generations* of living things. While the "bio" in astrobiology would make us think that this field would be deeply concerned with defining what an individual living thing is (so that we would know when we find it), astrobiology is actually more concerned with establishing an adequate general theory of living-systems. Astrobiology is dealing in populations. In a sense, we are searching for life as a planetary phenomenon on the level of a biosphere or microbial life teeming in a specific ecosystem on a habitable planetoid to such an extent that it may even seem appropriate to attribute the term "life" to the whole planet itself.[8] Technically put, we are searching for a widespread non-equilibrium system driven by a persistent geophysical or geochemical energy gradient.

Returning to Pilcher's definition specifically, to understand these generations of "life beyond Earth" we must also identify how the universe can "harbor" these living-systems. Life does not arise everywhere. Living things exist in a tight correlation with the environment in which they find themselves; changes to that environment can be lethal to a given set of living things. Astrobiology takes this connection radically seriously, and so is vested in better understanding the conditions of habitability. This emphasis on understanding habitable conditions clearly appears in the *Strategy* document as well, which identifies the "co-evolution of life and the physical environment"; "identifying, exploring, and characterizing environments for habitability and biosignatures"; and "constructing habitable worlds" as crucial emphases for future research.

There are at least four habitability requirements that need to be considered for understanding how the universe might harbor life beyond Earth: the availability of a liquid solvent, the existence of a corresponding chemistry, an energy source for metabolism, and a consideration of the breadth at which these other requirements pertain.[9] The first three of these are deeply interrelated: There needs to be a definite fittedness between the solvent, underlying chemistry, and available energy.

The liquid solvent localizes the region in which the chemical reactions of a living-system can occur. It provides a finite space for reactants to encounter one another and guarantees some consistency in the pressure and temperature at which those reactions may take place.[10] Any particular liquid solvent—usually water but other possibilities exist—could provide the medium for the self-sustaining chemical system of a given population of living things. However, not every chemical system will work in every sort of solvent. As any basic chemistry class indicates, understanding solvent effects is important. This is no less true in astrobiology. For instance, the methane lakes of Titan would not provide a great solvent for the organic chemistry attuned to water here on Earth; one would expect a different basis for any self-sustaining chemical system arising there.[11]

Whatever chemical system we might imagine to operate in a given solvent, the system needs to have sufficient flexibility to form a variety of stable polymers conducive to a broad network of enzymatic reactions. This makes possible the formation of a metabolism well suited to the energy sources available in the wider environment. Much like the solvent, not every network of enzymatic reactions is well suited to every sort of available free energy. If this seems complicated, simply remember that sunlight may provide abundant energy for plants, but it will not do anything for chemoautotrophs at Lost City. Just as there is a fittedness between the solvent and the underlying chemistry, there needs to be a certain fittedness between the available free energy and the network of chemical reactions taking place.

Moreover, those energy sources need to provide a minimum amount of free energy such that metabolic work can occur. Determining what the lowest threshold of such maintenance energy might be is very difficult because it relies on the particular properties of the solvent and chemical system in addition to wider environmental conditions such as temperature, pH, and salinity. Generally, though, the more abundant and accessible an energy source is determined to be in a given environment, the more likely it is that this environment could provide a habitat of astrobiological interest.[12]

The fourth habitability requirement—a consideration for the breadth at which the other requirements pertain—is in some ways the most interesting. What constitutes the habitability of a particular astrobiological phenomenon is not uniform. In the case of exoplanets, we are searching for continuous planetary habitability—where the other three conditions are being provided for at a geological time scale across significant portions of the surface of a planet. By contrast, with an icy moon like Enceladus, the habitability of interest is instantaneous—can its present conditions support a living-system in one particular sub-surface region. Mars provides an interesting balance of looking for geological evidence in the past of longer scale continuous planetary habitability and a contemporary focus far more specifically on instantaneous habitability.[13]

A broader picture emerges when astrobiology is a science that studies the harboring of life across the cosmos. Astrobiology is interested in where and how whole generations and populations of microbial living things might come into existence in the universe and proliferate into all sorts of complexities. Stretching to the far reaches of the cosmos, it is examining something intensely personal as well: It is a riff on that favorite human question, "Why do we exist?" What astrobiology makes abundantly clear, however, is that in order to understand how living-systems begin and develop, one must simultaneously analyze the planetary and environmental conditions that by the particularity of their habitable qualities shape the possibilities available to any given living-system. Living-systems are situated in wider habitable environmental conditions that in turn come to be shaped by those living-systems themselves. This is not revolutionary; biologists have known this for a long time. However, sufficiently understanding either living-systems or the habitable conditions for these living-systems in the astrobiological perspective requires understanding how the thoroughgoing mutuality of the relationship between them works. In astrobiology, one cannot study *either* living-systems *or* habitability; this is a both-and affair.

We need language that captures the extent of this mutuality. Various forms of "new materialism" are critical resources for this endeavor. Here, the concept of "intra-action" employed by Karen Barad's understanding of agential realism specifically provides a helpful conceptual tool. Working from problems of wave-particle duality in quantum physics to produce a conception of philosophical realism that is consistent across various scales (from the micro to the macro), Barad's account of "agential realism" makes the case for using flexible primordial units in her ontology. She calls this a phenomenon, though cautions it is not to be confused with the technical sense of a phenomenon in phenomenology;[14] Barad uses this term to refer to the fundamental wholeness of a measuring apparatus and the measured object. *Reality consists of phenomena* as the primordial units of her ontology.[15] More formally, "[T]he primary ontological unit is not independent objects with inherent boundaries and properties but rather phenomena. In my agential realist elaboration, phenomena do not merely mark the epistemological inseparability of observer and observed, or the results of measurements; rather, *phenomena are the ontological inseparability/entanglement of intra-acting 'agencies.'*"[16]

I simply want to introduce her neologism "intra-action" here as a helpful conceptual tool for astrobiology. In its most basic sense, Barad is using intra-action to note a conceptual shift from what *inter*action usually designates. *Inter*action assumes the prior existence of entities that relate to one another. *Intra*-action connotes the priority of the phenomenon as a holistic unit. As she variously describes it, intra-action indicates that "relata do not preexist relations"; or, that "[r]eality is composed not of things-in-themselves or things-behind-phenomena, but of things-in-phenomena."[17]

So phenomena are the basic ontological units of agential realism, and these phenomena consist of indeterminate intra-actions that under specific conditions give rise to what we think of as entities with determinate properties. The intra-action at work in phenomena is the meaningful context, which itself is the most basic ontological unit, in which what we traditionally think of as matter and material things take shape. As Barad makes clear from the very beginning of her work, meaning and matter are inseparable.[18]

Set in the context of astrobiology, I suggest that the term intra-action gets at the mutuality entailed between a living-system and the habitable environment. Rather than thinking about the intersection of these two as distinct phenomena *inter*acting (distinct parts of a greater whole), in astrobiology living-systems and environmental habitability form an *intra*-active phenomenon that is *ontologically primordial*. The meaningfulness of a living-system is not something that can be determined in contradistinction to its habitable environment: These concepts work in intra-active tandem—things-in-a-phenomenon—as a meaningful unit of existence that we cannot tear asunder without violating what constitutes a sufficient understanding of either part (the living-system or the habitable environment) in itself.

This is to apply Barad's term "intra-action" at a vast scale: the intra-action of astrobiology as it tentatively offers some rules for delineating the difference between living and non-living-systems. Taken as an intra-active discipline, astrobiology forces us to reframe what we mean when we use the adjective "living" as a descriptor. In so doing, life is better understood as a planetary or statistical quality of certain phenomena, not a descriptor of specific organisms. Bluntly put, the shift in our thinking would entail something like claiming it is not that a bacteria, bug, plant, or human being is alive; it is that the *systems* in which those creatures appear are *living* in a way that we might contrast with systems in which such features could not or do not appear that are non-living.

One of the greatest implications of understanding astrobiology this way is that it suggests the study of biology is moving away from being a descriptive science and toward a set of first principles.[19] Subsequently, we can then refigure how central symbols, like the *imago Dei* or other doctrinal loci, can continue to offer a meaningful lens on our existence when we understand that *human* being is an even more specific type of living-system in an astrobiological frame of reference. But this will be the focus of the next two chapters.

## A Persistence of Similarity and Usefulness

Before turning to issues of the *imago Dei* specifically, I want to return one more time to the four examples of astrobiological phenomena described in the previous chapter. It should be clear by this point that astrobiology understands the relationship between living-systems and the conditions of habitability to be fundamental to accounting for how life might exist *anywhere* in the universe. I am suggesting that this relationship is so crucial and thoroughgoing that it is intra-active, not merely interactive.

Nonetheless, considering the four examples of astrobiological phenomena discussed previously as intra-active should, at a minimum, make clear that astrobiology is not dealing with mere placeholders for an alien life-form yet to be discovered. Each phenomenon, in itself, is of direct astrobiological interest. There are some subtle differences as to how the interest and significance of each phenomenon is characterized based on its relative distance from the Earth and the detail of study for which this allows, as with the distinction between continuous planetary habitability and instantaneous habitability.

However, there seems to be one more unstated theme—beyond avoiding the abiotic, understanding living-systems, and conceptualizing habitability—emerging from these examples as well. There is a pernicious sense that establishing some thematic similarity to life on the Earth, no matter how near or far away the astrobiological phenomena might be, is crucial, or, it is at least crucial to the public presentation of findings.

We can think back to the two examples of exoplanets, in particular. In these cases, it is questions of habitability that take precedence because we are dealing with findings from indirect detection methods. No one has seen these planets—their existence and determinate qualities are estimates based on observations of the star that each orbits. The habitability thematized in these cases focuses on very basic criterion (if it is rocky, it could have liquid water and an atmosphere). In these cases, what is being sought is continuous planetary habitability on the surface because, practically speaking, there is probably no way in the near future we will be able to detect biosignatures on these worlds unless a fairly complex living-system is on the surface. Further hypotheses about the type of life that could be found there is then based on the wider properties of the stellar system in question. Again, the life that we might seek on these exoplanets may seem like individual organisms to the public, but this is not the case. Here astrobiologists are seeking life thematized as a comprehensive system: life understood at the level of the biosphere.

These efforts all fit with the themes developed in the previous section and indicate that the significance of studying these astrobiological phenomena relates to the value of understanding how a particular form of habitability relates to the emergence of a particular type of living-system. Yet journalists frame the public presentation of these findings in terms of a discourse of similarity or usefulness. Kepler-452b is similar because it circles a G2-type star. Proxima Centauri b is inviting because its star will last far longer than our own (in case we need to escape the Earth in a voyage of interstellar magnitude).

In the cases of Enceladus and Mars, we find a similar hidden appeal to similarity or usefulness. In the case of the icy moon, Enceladus, not only are themes about habitability and intra-action with an environment at play, there is also a shift in the level at which living-systems are being construed. No longer are we dealing with life as a property of a biosphere and issues of continuous planetary habitability. Instead, in the case of Enceladus we can make a much more specific analogy to *a particular group of thermophiles* that intra-act with an analogous environment of instantaneous habitability characterized by serpentinization *on our own planet*. Not only is the work being done related to Enceladus far more specific (which one can reasonably expect given the direct encounter), the rhetoric driving the description of habitability and the potential living-system rests on instantiating as much similarity as possible to a parallel system on Earth.

As was noted, Mars is a bit peculiar: It is of interest both historically and in the present in terms of astrobiology. The historical astrobiological significance of Mars parallels the significance of icy moons like Enceladus. The search for a fossilized record of biotic activity or fluvial conditions that might lead to likely locations for instantaneous habitability on Mars in the Noachian relies on rich comparison with specific similar features found on the early Earth.

The present astrobiological significance of Mars, however, is framed in ways most similar to how usefulness contextualizes the presentation of findings for Proxima Centauri b. Because Mars is so close that it could be reached, falls within the habitable zone of the Sun, and could be colonized—as Elon Musk has provocatively and perhaps infamously indicated—in conjunction with significant sorts of terraforming,[20] there is a working assumption of sufficient similarity and usefulness that conditions nearly all presentations of astrobiological research findings for Mars. In the case of Mars, similarity and usefulness not only frame findings for the public but also condition subsequent ethical questions. If Mars has microbes should we still go there? What would our future responsibility to another living biosphere be? Would the presence of subsurface microbes indicate we should keep our distance from the planet? These direct ethical issues are generally the types of questions that we immediately jump to when asked to consider the wider societal significance and implications of astrobiological research.

While neither the *Strategy* document objectives nor Pilcher's definition directly implicate similarity as a theme, nonetheless it is present in each of the examples. This is important to how we imagine the *transdisciplinary significance* of astrobiology. Kepler-452b is presented as a place you could get a tan; Proxima Centauri b is a first point to expanding into interstellar space; on Enceladus we might find lost cousins to life at Lost City; and the very fact that Mars is so near that we could likely go there within our children's lifetime makes the investigation into its conditions of habitability a simultaneous reconnaissance for finite Earth resources that could be mined as we exhaust the resources of our own planet.[21] Similarity to the Earth or usefulness for human beings becomes the rhetorical frame by which astrobiology is deemed significant for the wider public.

This raises two issues that need to be borne in mind as we proceed in imagining how astrobiology might contribute to shaping conceptions of the relationship between human beings and the wider environment. First, we need to be aware of how the rhetorical force of invoking similarity or usefulness bleeds into methodological suppositions for why particular courses of research in astrobiology should be pursued.

As a somewhat innocuous example, consider the nearly exclusive emphasis put on searching for water as the solvent in which living-systems can emerge. Previously, I framed this discussion, as is often the case, in terms of an extrapolation from known examples: We know of only one living-system (ours) that is based on using water as a solvent; other solvents may work, but for feasibility we should prioritize a search that makes use of the solvent that we know will work. The logic of this argument relies on similarity. However, one can continue to prioritize the search for water with a lesser used argument: Given the available elements and their frequency of occurrence in the context of cosmic evolution, water is likely to be the most commonly available solvent to support the formation of living-systems.[22]

The outcome of these arguments is the same in terms of research pursued, but the rhetoric used and what it implies about the significance of astrobiological research is quite different regarding how we invoke similarity to terran life. Essentially, usefulness or similarity may not be the only, or best, way to frame the transdisciplinary significance of astrobiology. This is particularly true if we can make statistical or probabilistic arguments instead of simple appeals to usefulness and similarity.

Second, we need to consider carefully how appealing to similarity or usefulness—particularly in the public presentation of findings—may inadvertently undercut the significant critique that astrobiology can suggest for imagining the relationship between human being and the environment. If astrobiology pervasively views living-systems and the habitable environment as an intra-acting phenomenon, this is significantly different from a more typical understanding of human beings as ontologically discrete subjects who happen to be situated in a given environment that is consumed or controlled in various ways to meet the needs of those subjects. If the significance of astrobiology is that it is moving us toward biological arguments based on first principles of living-systems instead of arguments based on descriptive catalogues of living things, this represents an important shift in the context of thinking about what constitutes meaningful human existence.

My hope for introducing intra-action is that it provides an alternative to this implicit appeal to "Earth-likeness" in how we frame the meaning of astrobiological discoveries that in turn provides a robust ground from which to make the case that understanding the intra-action between human being and the wider environment is necessary to living meaningfully. If intra-action is essential to understanding living phenomena, then an adequate account of meaningful human existence must take this ontological frame of reference into account. How we, as individuals might come to appreciate this frame of reference, and feel it as being of concern to us, will have to be considered more explicitly in Chapter 7.

Instead, the goal here has been to make a minimal and more preliminary claim: Astrobiology is a legitimate (and critical) context for engaging in correlative theological thinking. Not only does it invoke a concern for intra-action, but it does so across a vastness of scale that is incomprehensible to almost any other field of study for engaging this issue. If we take astrobiology, not in its popular conception as a placeholder science seeking the discovery of alien lifeforms, but as an interdisciplinary field studying what might constitute the first principles for the intra-action of living-systems with specific sets of habitable conditions across the entirety of the universe, it has a profound significance.

It articulates a harmonious cosmism that does not drive us away from Earth with some sort of utopic hopes of cosmic escapism or teleological inevitability of human ascent to planetary control.[23] Instead, astrobiology radically drives us back to the particularity of our existence on *this* planet in all its irreducible complexity. We are well-suited to live here on Earth; we are well-suited to this tiny rock situated in the vast, inky blackness of seemingly infinite cosmos. We are so well-suited to it that, perhaps, we cannot even realize what it would mean to be human apart from this rock that is our home. To articulate what it means for us to share in meaningful human existence anymore, to be the *imago Dei*, we must begin to think astrobiologically.

We need to reclaim symbols in new ways that will be cogent with this widened astrobiological context. It is fine to acknowledge that we want to know more about how the intra-actions of living-systems and habitable environments occur across the universe, but without symbols that engage this knowledge and help us *understand* how adopting this view meaningfully orients us to our reality in new and significant ways, then astrobiology remains a bit of a novelty. Moreover, whatever symbols are engaged for this task—even doctrinal symbols such as the *imago Dei* to which we will turn next—cannot be of significance *only* to closed, specifically delimited communities. Given the universe (literally) of applicability that astrobiology helps us imagine, the symbols that frame our meaningful human existence must always be seeking to develop values that invite the inclusion of broader publics.

# CHAPTER

# 3

# Being a Living-System

When my son, Henry, was five years old, we were playing catch with his new baseball glove in our front yard. It was early fall, and the maple tree's leaves had just turned brilliant shades of red and orange. After my third throw went gently over the tip of his glove and rolled behind him to the hedge, Henry ran to get the ball. As he passed the tree, his left foot caught on a root, and he fell to the ground in a cartoonish splay for which children seem to have a special knack. He popped up immediately, and I was both relieved and a bit amused. Henry, however, was angry. He promptly walked up to the maple tree and kicked it hard.

When I asked him why he would do this, he looked a bit incredulous that I did not have more sympathy for his plight: "The tree reached out and tripped me; it deserved that kick."

As a professor, this seemed like an opportune time to launch into a diatribe about the importance of trees to our ecosystem and how without trees we would not have oxygen in our atmosphere for all the complex animals on Earth to exist—even you and me. Henry was unimpressed: "Dad, it was just a tree . . . I didn't hurt it, it hurt me."

Revealing where Henry received at least some of his genetic predisposition to stubbornness, I asked him, "If your friend had pushed you down, would you jump up and kick him?"

"No!" he quickly replied.

"Why would you kick the tree, then?"

He looked at me as if I had rocks in my head and said, "Dad it's just a tree, not another kid like me. I didn't want to hurt it."

Abstract arguments with five-year-olds are recipes for tears. We went back to playing catch, further away from that maple tree, and planted some tulip bulbs around the tree so that the next spring we could remember not to get too close to the roots. Henry and I agreed that might be better for everyone.

In the last chapter, I identified three themes of astrobiology, but thematizing the indispensable intra-action of two themes was most crucial: the intra-action of a living-system and the habitable environment. These themes and their intra-action are axiomatic to astrobiology as a field; yet they are, at least to some degree, at odds with our everyday ways of encountering the world around us. Henry had no sense of being in intra-action with the maple tree, and it was clear to him that the maple tree was somehow inherently due less respect than another child. Even though he did not intend to hurt the tree, lashing out at it seemed naturally more acceptable. Implicitly, he had already developed a sense that the wider environment around him (like that maple tree) was there for his use and he could be angry with it when it stymied his efforts to get the ball so he could keep playing and having fun. Implicit to his encounter with the maple tree was a sense of human exceptionalism (a very minor exceptionalism—but exceptionalism nonetheless) that is at odds with the intra-action that grounds an astrobiological perspective on our existence. The intra-action of the astrobiological perspective cuts against appeals to human exceptionalism that early in our development becomes a "natural" way of perceiving the world.

If astrobiology provides a credible way of thinking about what it means to "live" on and with the wider habitats of *any* cosmic body, then it behooves us to consider how we can reinterpret existing sets of symbols in order to develop a meaningful way of being-in-the-world and belonging-together-with-the-world in light of the astrobiological concern for intra-action that counters tendencies to human exceptionalism. To do this, I suggest the astrobiological intra-action of living-systems and habitable environments should form the context of engagement for interpreting symbols that facilitate the process of meaningfully ordering our way of existing in the world. Thus, the next four chapters will each interpret one of two symbols, which I will argue can be corroborative, in light of the astrobiological context of engagement: the *imago Dei* (in Chapters 3 and 4) and the Anthropocene (in Chapters 5 and 6). How do we conceptualize being the image of God in light of this intra-action that is indispensable to astrobiology? And how do we conceptualize the meaningful order of existence the Anthropocene summons us toward in light of astrobiology's intra-action?

Choosing to deal with the *imago Dei* is hardly accidental. Doctrinally, it has typically been linked with arguments about human exceptionalism. If there is a symbol most likely to conflict with the way astrobiology has been framed here, this might be it. Our natural tendency might be to let this symbol die. Instead, I want to make the case that if we look into the location of this doctrine in cosmogony, then we might actually find a potent source of reflection, not for arguments concerning human exceptionalism as applied primarily to individuals but for a robust account of a responsibility to facilitate intra-action.

In this chapter, I will make overlapping arguments that deal with the nature of symbols. I will first examine the relationship between Christian doctrine and symbols to make a case for why doctrines might be reclaimed as symbols in constructive theological reflection if they are not used primarily for apologetic purposes but to facilitate the meaningful re-orientation of our existence in the world. I will then consider the *imago Dei* as such a doctrinal symbol. Resisting the tendency to turn the doctrine into a freestanding account of biblical anthropology, we find a resonance between astrobiology's account of intra-action and the harmonious ordering of creation that can ground subsequent interpretation of the symbol.

## Symbol and Doctrine

In a highly confessional or apologetic framework, a theological doctrine is a tradition or belief that aptly summarizes some locus of theological truth. The theologian systematically develops a doctrine in order to better defend its claims to theological truth by being situated in a coherent worldview that resists critiques levied from outside the system, and to provide a criterion for reasonably interpreting other doctrines within the system. For an apologetic approach to the doctrine of the *imago Dei*, specifically, what is at stake is establishing a theological truth about the meaning and significance of human being in contrast to competing secular claims that can then condition systematic reflection about other issues of interest with regard to human being within a given confessional community.

I want to move away from a traditional, apologetic approach to doctrine, following, instead, a trajectory of thinking about doctrine as a form of symbolic engagement. What one would traditionally consider a doctrine, like the *imago Dei*, here represents a specific form of *symbolic engagement with ultimate reality*. Before launching into an interpretation of the *imago Dei*, it behooves us to think about *how* this symbolic engagement works. As Robert Neville keenly indicates, "[D]octrines by themselves are neither true nor false. Only when they are asserted in an interpretive act of some sort (many sorts exist) might they be true or false."[1]

Hopefully, using the language of "symbol" and the "ultimate" draws a clear allusion to the work of Paul Tillich for the reader. It is his account of a symbol that gives initial direction to the trajectory of my thinking.[2] Tillich asserted that a symbol is different from a sign by participating in whatever it points to. Simply put, symbols are fragments of finite reality that participate in a transcendent, infinite ultimacy. This participation, crucially, creates an inescapable inner tension to symbols. Symbols participate in ultimacy, but that participation is only manifest insofar as the symbol is "of concern" to our concrete existence. Here is where the tension arises. The index of our concern, our ability to be concerned, is in direct correlation to the concreteness of the object of our concern. For a universal concept to be of concern at all requires that it be represented through finite, concrete experiences. In contrast, for something to be truly ultimate, it must transcend everything finite and concrete. As this transcendence occurs, however, that which is ultimate becomes increasingly abstract.[3]

This is the inescapable inner tension of our being ultimately concerned, and it is a tension imbued upon every symbol that facilitates our participation in ultimacy. Symbols make the ultimate available through some particular feature of the symbol that is of concern to us, wherein what it means to be "of concern" is that a symbol makes a claim upon how to shape our own meaningful, concrete existence. By their very nature, symbols make a normative claim upon us. They summon us to a particular way of being in the world with others that then constrains the ways we might live humanly together.[4]

I want to stress, though, that for a symbol to be meaningful to a given individual, for its concern to register with existential import and make a normative claim upon us, we must interpret it. Such an act of interpretation always occurs in a specific *context of engagement* that provides the hermeneutical framework in which this act of meaningful interpretation can take place. The context of engagement matters because it conditions precisely how the symbol appropriately communicates its participation in ultimacy to an interpreter (or even if this communication is possible).

The "appropriateness" is determined in relation to how well the ultimacy the symbol makes available is able to provide a sense of fulfillment or flourishing that is promised when proximate concerns are subjected to this ultimate concern.[5] One cannot simply divorce the symbol from the hermeneutical act of its interpretation, as though the ultimacy it communicates, even if normative, is statically available. This is, at least in part, because symbols summon us to a particular way of being in the world with others that then constrains the ways we might live humanly together. They speak to a primordial sphere of relationality—an interhuman depth that arises in deciding how we might be with the vulnerable face of the Other—that is persistently dynamic.[6]

Edward Farley helpfully extends this discussion in an additional, and important, way. He indicates symbols are "the values by which a community understands itself, from which it takes its aims, and to which it appeals as canons of cultural criticism."[7] This disrupts any lingering personalistic individualism that might errantly shape our understanding of symbols. While individuals meaningfully take up symbols, they are squarely rooted in communities.

These communities give shape and initial meaning to symbols by orienting them in wider associative universes—using *sets* of symbols to condition how any *particular* symbol might participate in ultimacy.[8] Thus, we cannot simply create a symbol *de novo* or ignore it entirely without significant consequences. The meaningful ordering of our reality as it is conjunctive to a sense of ultimacy relies on the shifting interdependences of symbols in these networks.

One of the crucial consequences of treating doctrines as symbols in this way is that we must always render them as *fallible*. Symbols arise in the context of particular historical and geographic settings. When too tightly tied to these originating source communities and their locales, symbols can become unmoored from the questions and existential realities that drive our current search for meaningful forms of existence. In this case, a symbol can be false because it is simply irrelevant.[9] Vice versa, even though symbols arise in particular historical and geographic contexts, they can easily be absolutized. The connection to ultimacy and meaningful existence they open can cause them to become fixed so that they are assumed to have an idealized significance across all historical epochs and geographic boundaries. In this case, symbols that were once life-giving means of participating in ultimacy can become demonic idols—disenfranchising persons or groups by uncritically adopting the symbol.[10] If a symbol or doctrine is to help us flourish, we must be vigilant to reinterpret it critically with an eye for its relevance in shifting contexts of engagement. We must always consider how our interpretations of symbols address the historical relativity and the potential subjection of a symbol, particularly a doctrinal symbol, to idolization.

To summarize the argument so far, I would highlight three ideas. First, a thorough understanding of any symbol requires we pay attention to the particular features of the current context of engagement in which it is being interpreted. Second, we must give careful consideration to the communal history and network of symbols that have facilitated the interpretation of any given symbol that lead up to its meaningful adoption in the current context of engagement. Third, the power of any given symbol as interpreted in light of a specific context of engagement depends upon its ability to facilitate our flourishing by promising participation in ultimacy through its concreteness.

These three points are particularly important when we consider *doctrines* as symbols. On the model outlined here, as opposed to a purely apologetic approach, any doctrine should be judged based on its ability to function symbolically. In a given context of engagement, does the doctrine facilitate our participation in ultimacy by giving some new lens on the concrete world? The primary function of a doctrine, then, is not to look inward to the systematic development of a comprehensive worldview or merely to defend the cogency of one worldview in relation to others. Instead, doctrines as symbols ought to promise fulfillment for anyone situated in the context of engagement from which the act of interpretation occurs. A doctrinal symbol has to be considered in light of a widened context of engagement; it needs to be interpreted with respect for how it opens up an existentially meaningful encounter with ultimacy *regardless* of whether that person is considered part of the community from which the symbol originates.

If the context of engagement is suitably narrow (for instance, persons within a specific confessional community), then the apologetic approach to doctrine might remain unchanged. However, if the context of engagement being considered is much wider (as with many forms of constructive, public, and political theology), then our consideration of doctrine will be quite different. It is tempting to simply throw out doctrines in this latter case: to treat doctrines as dead symbols that can no longer speak to our most significant, broadened contexts of engagement. Yet this is a move to which we rush too quickly. Symbols, such as doctrines, that have stood the test of time do not die so easily and are often quite flexible if we look deeply into their development. They speak to wider publics and these wider publics affect the ongoing interpretation of a doctrinal symbol.

If we build on this flexibility, then I think we can, and should, reclaim doctrines insofar as they facilitate the development of transreligious, public values, values that speak to the needs of the wider fabric of society and help develop a sense of our responsibility and accountability to all kinds of planetary others (human and non-human). What I mean is that when a doctrine is able to promise fulfillment (broadly construed as a meaningful way of existence ordered to some sense of ultimacy) when interpreted in a broadened context of engagement, especially for those not participating in the community of origin for a given doctrinal symbol, then these doctrinal symbols have the power to transform our wider society.

This is a power of doctrinal symbols that should not be cast aside lightly. Instead, we should look for the ways in which doctrinal symbols may meaningfully construe existence in light of broadened contexts of engagement and how this might contribute to promoting a more just society. In short, I do not want us to throw out doctrines just yet. If they are interpreted as symbols in which a meaningful sense of ordering our reality toward ultimacy is established in light of contexts of engagement with significance for outward-facing, wider, pluralistic publics instead of closed confessional communities, our doctrines may have new life in various forms of constructive theology.

## The *imago Dei* as Doctrinal Symbol in an Astrobiological Context

What I have claimed so far about doctrinal symbols generally, I will now consider more specifically in light of the *imago Dei* and astrobiology as the context of engagement for interpreting this symbol. It is worth noting from the outset, however, that there is something peculiar about this pairing of the *imago Dei* and astrobiology. Astrobiology is a *distinctive* context of engagement that we cannot simply equate with how existing discourses in theology and science related to the *imago Dei* have proceeded. The approaches taken in theology and science discourse have given far more attention to other fields of science.

For instance, we find a tendency toward a nearly ubiquitous typology that organizes the historical treatments of the doctrine within the subfield of theology and science.[11] The typology employs three divisions for interpretation: substantial, relational, and functional. There is overlap between these types, but this schema of categorization has become a common fixture in secondary literature. Roughly put, the substantial focuses on characterizing human uniqueness in terms of a distinctive faculty (most often reason); the relational characterizes human uniqueness in terms of the capacity for relationship (either between human beings or between human beings and God); and the functional focuses on the biblical injunction to care for creation in God's stead (as though we are a proximate gardener for Eden).

The ubiquity of the typology should not be altogether surprising. Even when the *imago Dei* has been considered in some broader fashion, such accounts have at least *implied* that the significance of this doctrine is connected to defining what distinguishes us from the rest of creation. The typology's focus on uniqueness or distinctiveness indicates this well. Moreover, the typology provides an exceptionally helpful heuristic for historical theology (one that does not over determine how any individual theologian's account of the doctrine is shaped into a broader, systematically consistent theological anthropology). It represents an accessible classificatory schema that leaves wide latitude for developing the implications of each type in diverse ways such that the doctrine quite readily remains systematically consistent with other features of theological anthropology (for example, a doctrine of sin, theological ethics, soteriology, and so on).

The typology has been important for research in science and theology not only because of its heuristic value, but because it can be readily mapped to means of considering human distinctiveness in physical anthropology and human evolution. It correlates well to a Darwinian search for the evolutionary distinctiveness of human being. As Wentzel van Huyssteen has intensively demonstrated, there is a productive transversal intersection between this doctrine and the consideration of features of human distinctiveness in physical anthropology and studies of human evolution.[12] Given the heuristic value of the typology and its ready pairing with a narrative pursuing the evolutionarily distinctive features of human being, it should come as no surprise that interdisciplinary research in theology and science concerned with the *imago Dei* has given special, if not nearly exclusive, attention to findings in physical anthropology and the study of human evolution. Constructive accounts of the *imago Dei* in theology and science focus primarily on how individual human beings possess or come to exemplify evolved sets of traits, capacities, or skills.

Of course, anything so easily categorized is also easily troubled. Challenges to this standard typology and questions about its effectiveness have become nearly as ubiquitous as employment of the typology itself. Criticism of the typology has suggested that (1) it erases the complexity of individual theologians' reflections on these issues; (2) the doctrine lacks any strong sense of a fixed meaning despite its wide use because it serves as a catch-all for overlapping questions about theological anthropology in general—wherein the questions to be emphasized depend on the needs of a given context; and (3) the various arms of the typology work on a self-undermining logic that forms a bad dialectic.

While the first critique is simply *pro forma* for any typology, the second and third critiques are more substantial. The second critique reframes the advantageous latitude—allowing for flexibility in the systematic coordination of the doctrine with other facets of theological anthropology—as a weakness. Olli-Pekka Vainio contends the lack of a fixed meaning for the *imago Dei* simply indicates that the doctrine never became a contentious, central topic of theological debate: There is latitude because it is of marginal importance. He identifies seven relevant questions to which the *imago Dei* has pertained at various points in history and suggests we now need more holistic understandings of the doctrine that incorporate these diverse questions without unnecessarily pitting them against one another—as with simple approaches to the typological view.[13]

This call for a more holistic understanding moves us to the critical features of the third critique. The self-undermining structure of the typology is easy to rehearse. Each category in the typology makes its claim to distinctiveness by sublating the other categories—reducing claims that might fall within one of the other categories to the orienting assumptions of the given predominant category. In the case of the substantial arm of the typology, it is easy to imagine that claims from the relational or functional categories are merely veneers for a faculty or host of faculties that are reducible to substantial properties possessed by a given individual. In the case of the relational, the various features of the substantial and functional categories are merely individuated properties inappropriately divorced from the richer relational context. In the case of the functional, the substantial and the relational categories are merely byproducts of enacting a divine command to care for the rest of creation that gives us meaning and purpose.

Our initial impulse might be to challenge these reductive appeals to enfold the other categories into a singular, dominating aspect of the typology directly, thereby hoping to reinstate some balance between the categories. The vaunted reason of the substantial approach is not as distinctive as might be hoped—as various cases in animal intelligence studies demonstrate. A thorough-going ontology of relationality could easily swallow human distinctiveness and individuality entirely if logically extended far enough. Finally, the problems with a reductive appeal to the functional approach, somewhat ironically given it is often most lauded by advocates of ecotheologies, is that it is wrapped up with a specific socio-political schema at work in the cultures from which it arises that can easily be used to invoke a standpoint of cool self-removal from which dominion is to be enacted. Yet these direct challenges seem only to lead to further quibbling through various acrobatic acts of exegesis and *ad hoc* justifications intended to elevate once again a predominant category.[14]

It seems there is a more pernicious problem here. I would suggest this logical ouroboros indicates that the familiar typology for the *imago Dei* is quite limited in its explanatory power, particularly insofar as it leads us to assume that human uniqueness or distinctiveness is a theological *problem to be solved* by appeal to some simplest quality. The *imago Dei* becomes merely stipulative: concerned with designating some appropriate condition for drawing a line in the sand that puts human beings on one side and the rest of creation on the other.[15] All too quickly, then, this approach becomes a means of pursuing a fetishistic separation of human being from the rest of creation (i.e., we do something no other living things do). We rest on the top of, or at a pivotal point within, a simplified chain of being—regardless of whether we understand this apogee in terms of some ability to reason, relate, or care.

Moreover, the focus on uniqueness or distinctiveness as the locus of concern for interpreting the *imago Dei* is problematic if astrobiology provides the context of engagement for interpretation. If we fixate on stipulating uniqueness, it is hard to see how engaging astrobiology does anything besides play out the same questions driving the existing cross-disciplinary engagement with physical anthropology and evolutionary biology only at a different scale. What theologians once considered on a smaller, planetary stage is now set in the context of a bigger, cosmic stage: At issue is how we find some way to hold onto our sense of distinctiveness in the context of the universe, not just the planet. Colloquially, how do we remain a big fish in an ocean instead of just a small pond?[16]

By contrast, though, I am suggesting astrobiology provides a legitimate context for meaningfully interpreting and engaging with the *imago Dei* that shifts the way we would interpret how this symbol participates in ultimacy and is of concern to us. Using astrobiology as the context of engagement for interpreting the *imago Dei* does not just change the scale at which the usual discourse about human distinctiveness plays itself out. The significance of the differences emerges when we return to the key concept of understanding the intra-action of living-systems and the habitable environment.

With regard to the origin and development of living-systems, the type of living that astrobiology describes is distinctive. As was already emphasized in the previous chapter, a quotidian approach to living things brings individual organisms or small groups of organisms to mind. You might think of other people in a city, birds in a forest, a group of insects in leaves, or fields of tall grasses swaying in a breeze. Perhaps some image from a microscope slide comes to mind teeming with chlamydomonas and paramecium. In any case, our commonsense approach to this issue is to think of individual organisms of some sort, perhaps in a group.

Yet astrobiology is interested in living as a material and organizational system. Astrobiology (particularly in the way that I have been presenting it here) is not, strictly speaking, interested in the individual organisms that so readily come to mind. The individual organism is of interest insofar as it contributes to the process of shaping the intra-action of a living-system with its habitable environment. In a sweeping way, we could say that astrobiology is more concerned with life in its systemic generality across generations than in its organismic particularity. Part of our focus in this chapter must be on offering an account of the *imago Dei* that takes up this concern for systemic generality instead of organismic particularity. The *imago Dei* must be figured as "ours" not as "mine."

With regard to habitability, this means emphasizing that human beings—just like all living things—are situated in a particular habitable environment. Moreover, living things intra-act with this environment. What astrobiology emphasizes is that living things always find themselves thrown into a habitable environment and the habitable environment is habitable only insofar as it does or could support the existence of living things. This is a relational system that we cannot assume is best characterized as being composed of "living things" and "the habitable environment" as discrete, separable entities.

If we take seriously how astrobiology has something profound and distinctive to say about the two crucial themes orienting it, then what we might claim more generally about the intra-action of simple living-systems and the habitable environment ought also to apply to how we understand human being insofar as it, too, is a living-system. Human being can certainly be different or more complex, but whatever we might claim, theologically or otherwise, about what it means to be a human being cannot be at odds with axioms explicating how life can emerge across the universe. Any account we offer of human being has to be sensitive to the broader ways in which it represents a particular sort of intra-action between a specific living-system and the coordinate habitable environment.[17]

Given these astrobiological challenges to the *usual* way of setting up the *imago Dei* story, we need an alternative approach that helps avoid the pitfalls outlined previously. Of course (given the breadth of questions to which this doctrine refers and the disagreement of theologians over how one might categorize attempts to make use of it), we could simply abandon the *imago Dei*. I would suggest, however, that the *imago Dei* still serves as a powerful symbol to help us interpret our existential reality and orient our experiences of ultimacy in the midst of that reality. In particular, it is a place from which theologians can enter into a wider transdisciplinary conversation about what it means to be human in the midst of our world and how that understanding of humanity speaks to the ways we frame a sense of well-being. Let's not throw it away just yet.

## A Creature of Cosmogony

To do this, however, we need to take seriously that the most theologically significant texts dealing with the concept of the *imago Dei* in the Hebrew Bible have been three selections from the primeval history of Genesis—1:26–27; 5:1–3; and 9:6. All three instances come from the priestly narrative, and traditionally it is the first text, 1:26–27, on which the latter two rely. These latter selections are references within the Noahic story that serve as literary allusions to the first creation story: that humankind, both male and female, are the "image" and "likeness" of God—ṣelem and dəmût.[18]

We will look more closely at the semantic range and theological significance of these lexemes shortly, but first we must situate understandings of the *imago Dei* in the wider genre of biblical cosmogonies. We cannot overlook that to be the *imago Dei* is explicitly to be a human *creature*: part of the wider cosmic ordering characteristic of cosmogony. While in the context of the opening cosmogony of Genesis 1:1–2:4[a] human beings appear to be peculiar creatures placed intentionally in a different orientation to other creatures, this peculiarity of human being hardly appears to be a fixture of the genre.[19] If we bear in mind this wider cosmogonic literary context and its theological implications in interpreting the *imago Dei*, then we must emphasize three facets of how biblical cosmogony attests the well-ordered shape of reality: meaning, consistency, and well-being.[20]

First, the divine ordering of creation is extensive. All things that exist are a product of the creative power of God, and this divine act is something otherwise unheard of in its newness.[21] Creation is a field contrasted to an initial chaos from which meaningful existence is not self-derived.[22] From light to Leviathan, to be *something*—to exist as a particular determinate thing—is to have come about under the auspices of a divine creative power that resists the threatening devastation of meaninglessness. The act of creation itself is not a mere calling forth of determinate things into existence over and against chaos: an origin story emphasizing the radical distance and transcendence of an orderly creator.[23] Creation is a transformation of what possibilities open to determinate creatures. It is an ordering of creation as the opening of complexity and possibility—possibilities that push beyond banal, bare existence toward a meaningful existence for *all* creatures. To be the *imago Dei* is to fit into this wider schema of a creaturely possibility for meaningfulness.

Second, the plentitude God brings to be occurs with a reliable consistency. If the act of creation is about possibilizing meaning making, the ordering that occurs throughout cosmogonies suggests that the consistency of a persistent sense of order is coordinate to actualizing this possibility of meaning making. A sense of order and expectation by which the needs of a creature can be satisfied and attained is laid out for all the elements of this creative plentitude while simultaneously instantiating a dialectical tension that needs to be acknowledged: a tension between the creator and creatures. God is indeterminate (infinite) while the rest of creation is determinate (finite). Consequently, whatever we might want to say about God, such a statement will always remain incomplete; God is an inexhaustible mystery that persistently exceeds the whole plentitude of creation. The depiction of God from cosmogony is as always beyond determinacy and even abysmal—"God beyond god" or "God as being-itself not a being."[24]

Finally, this creative act, full of meaningful ordering and established consistency, is not ambivalent in any sense. The ordering provides a means of meeting the needs of creatures such that meaningful existence is possible. Concomitance to this order becomes indicative of a state of well-being for creation in which its possibility of meaningful existence can be realized.

Two key points follow. First, the conditions for realizing this possibility of meaningful existence are built into the structure of cosmogonies: the creative act not only makes possible meaningful existence but is also ordered toward the realization of this possibility. Existence as creation has structured intention; being is realized not as banal materiality but the well-being by which the possibilities of meaningfulness can be fulfilled.

Second, this ordering and its concomitant sense of well-being for creation *cannot* be divorced from the act of creation itself: To create *is* to establish well-ordered conditions that meet the needs of creatures in order to pursue meaningful existence. God as *prius* cannot be understood in contradiction to this creative action; the creative act (with all the well-ordered care and intention cosmogony intends to demonstrate) is never other than the ultimacy from which it precedes. To borrow an idea from phenomenology, there is an "aboutness" to the meaningful structure of existence as creation that is linked to the intentionality of the creative power of the divine. God's creative power—fundamentally giving rise to the well-ordered creation of cosmogonies exhibiting a sense of well-being—indicates that graciousness and delight constitute God's "transforming disposition" toward the world.[25]

Resultantly, cosmogony assures us that even as God is beyond all structures of existence, God who is being-itself or the power of being is not ambivalent to the needs of creatures. Again, the existence that is the product of God's creative act is never sheer existence, but meaningful existence. Existence exhibits the attentiveness not of banal being but well-being. Such an assurance calls out for a human response that becomes evidenced by the poetic lyricism of cosmogony itself and a desire for a living out of this order—of realizing new possibilities of meaningfulness—in unbroken response to the transforming disposition revealed in God's creative acts.[26]

What we should take away from this brief review is, first, that these cosmogonies represent a remarkably more sophisticated view of the world than a cursory glance might suggest. Moreover, it is a view that in its poetic and theological qualities is certainly coherent with the intra-action stressed by astrobiology. The cosmogonic interplay of creatures living out the possibilities of meaningful existence as part of a harmonious order attuned to their flourishing indicates how readily astrobiology can be a context of engagement (given a certain shared affinity with the worldview at work in cosmogonies) for interpreting the *imago Dei* if that doctrine is not removed from this wider context.[27]

If being the image of God is interpreted within this well-ordered, wider context of cosmogony, the *imago Dei* should describe how we, too, enact possibilities of meaningful existence that are rooted in our very creatureliness. This is a far cry from separating human being from the rest of creation. However, this does require a certain resistance to cleaving away the *imago Dei* from its wider narrative and forming it into a freestanding statement about biblical anthropology. Nor can its implications for theological anthropology simply be subverted within a wider doctrine of creation.[28] Instead, we need to perpetually consider how the harmony, order, and consistency of such cosmogony calls out for celebration that requires a poetic and aesthetic sensibility (one evidenced by the very creation of cosmogonies) to communicate the awe and wonder it inspires such that the image of God extends the creative ordering and possibilizing of meaningful existence.[29]

If we take the cosmogonic location of the doctrine of the *imago Dei* seriously, then astrobiology provides a viable and significant context for interpreting this doctrinal symbol. It is a context that is quite different from preceding accounts of the *imago Dei* in theology and science that have situated this doctrine more directly in physical anthropology, human evolution, and a quest for an account of human distinctiveness. Instead, emphasizing the work of God in creation to ensure our meaningful existence, not simply bare existence, the *imago Dei*, in an astrobiological context, should include at least two facets.

First, being the *imago Dei* should describe a particular type of intra-action between a living-system and the habitable environment. To be the *imago Dei* is to be created as a living thing. If we take the astrobiological impulse to think of life in terms of the intra-action of living-systems and the wider habitable environment, then the *imago Dei*—in this context of engagement—would be represented as a particular type of intra-action.

Second, the intra-action of astrobiology describes a certain sense of fittedness between a living-system and the wider habitable environment; this fittedness is echoed in the cosmogonies from which the doctrine of the *imago Dei* is drawn as a promise in creation. Creating is not bare existence, but a form of meaningful existence: creatureliness with the promise of flourishing. However we might interpret the *imago Dei* in light of astrobiology, it must connect to this wider promise of flourishing that is central to biblical cosmogonies and seems to metaphorically overlap with astrobiology's concern for fruitful intra-action.

Keeping this ground in mind, we must now briefly turn to consider key ideas in the longer history of the *imago Dei*. This will allow me to illustrate just how flexible this doctrine might be and what the significance of re-interpreting the doctrine in light of astrobiology is, especially with an eye to its significance for wider publics and not merely confessional communities.

# CHAPTER

# 4

# The *imago Dei* as a Refractive Symbol

The last chapter emphasized that symbols have a history. We cannot simply make them anew to suit our whims or change them arbitrarily. Here, I will provide a selective history that emphasizes how the *imago Dei* has not had a fixed meaning but, at different times in Christian history, has been associated with a host of other doctrines—such as creation, the trinity, and christology. While the historical work in this chapter may seem like it locates the significance of the *imago Dei* within certain confessional Christian communities in an apologetic fashion, this is not the intent. Instead, I want to better understand how this history of interpretation and association with other symbols demonstrates the flexibility of the *imago Dei* to make our participation in ultimacy "of concern" in diverse ways. We must be clear about what we can draw forward—and what will need to be left behind—from this history in order to offer a rich, constructive account of how to interpret the *imago Dei* participating in ultimacy in light of an astrobiological context of engagement.

What follows is not by any means a comprehensive history of the doctrine of the *imago Dei*. Instead, this is an effort to better understand *what* we have claimed historically, *why* these claims had and continue to have existential significance, and *how* inescapable themes of the *imago Dei* have been re-interpreted at different points in history so as to be relevant to the various contexts of engagement that previous theologians have sought to address. Thus, this brief history forms a ground for constructively imagining what the *imago Dei* might mean if it is to speak in a relevant way to a contemporary sense of astrobiologically existential significance.

In what follows, I will focus on two key themes that constructive accounts of the *imago Dei* must address: the continuing relevance of the image/likeness distinction beyond its original exegetical framing and how what we mean by "image" might be better theologically rendered as "symbol." I will pursue both concepts with an eye to situating the doctrine in the wider biblical cosmogony from which it arises while focusing on three historical theologians: Irenaeus, Augustine, and Schleiermacher. The work of each has driven forward these crucial themes in distinct ways.

Irenaeus's account of the image/likeness distinction, while exegetically untenable, crucially complicates what it means that we neither are devoid of the *imago Dei* nor perfectly attuned to it. Augustine takes the history of the doctrine in a new direction by not focusing on Irenaeus's fundamental distinction and instead imagining how being the *imago Dei* is a point of symbolic participation in the divine life that opens up the Trinitarian life of God in ways otherwise unknowable. Finally, Schleiermacher provides a first glimpse of what symbolic participation in the divine might mean as the *imago Dei* if we do not make recourse to the metaphysical assumptions and narrative of salvation history that takes an account of the Fall as a historical event.

From these few seminal moments in the history of doctrinal reflection on the *imago Dei*, I am building a case for what constitute inescapable elements of this symbol. From this historical recapitulation, I argue that to be the image of God is to be a *symbol* of God: one who refracts the creative power of God evidenced in cosmogonies to facilitate the flourishing intra-action of living-systems with the habitable environment. The consequence of this approach is that to be the *imago Dei* is not something properly ascribed to any individual organism as a marker of distinctiveness, but it describes a particular type of astrobiological intra-action.

## Image and Likeness

Before analyzing the contribution of Irenaeus, it is worth returning one more time to the primary textual references that give warrant to this doctrine. Of the three references to the *imago Dei* in the primeval history of Genesis, the first two occurrences, 1:26–27 and 5:3, each make use of both of the key substantive terms associated with this doctrine—ṣelem and dəmût—while the final occurrence

in 9:6 only uses the word for image, ṣelem. Of the two terms, dəmût
(traditionally meaning "likeness") is more common. Often translated as "like" or "as," it is a general term of comparison, though
more specifically it denotes similarity in appearance or form. It is
in deference to the semantic range of the term that a contemporary
translation, like that of the Common English Bible, will even render
dəmût as "resemblance."

By contrast, the word ṣelem—traditionally meaning "image"—
appears less frequently: seventeen times in the Hebrew Bible, as well
as seventeen times in Biblical Aramaic.[1] While the Proto-Semitic
root *ṣlm never appears as a verbal form in Hebrew, it has numerous
references in the verbal development of other Semitic languages,
perhaps most notably Arabic, where the cognate verb means "to
chop off, hew, cut, or carve." From the Arabic verb we might infer
that the Proto-Semitic verbal root would probably have been used to
describe a craftsman's or artist's process of creating a tool, patterned
implement, or sculpture; the substantive instances of "image" or
"idol" related to this root appearing in various other Semitic languages supports such a connection to the verb.[2]

While the term ṣelem is far from being a *hapax legomenon*, there
remains a curious separation of the meaning of ṣelem in Genesis
from its occurrence elsewhere.[3] It seems likely that the partition-
ing of these three specific instances from other occurrences is in
part due to the influence of classical accounts of the doctrine of the
*imago Dei*. Such accounts often relied on positing an exegetical distinction between ṣelem and dəmût in these instances where the two
terms appeared in succession (thus noting where such successive
appearances occurred was important).

This exegetical distinction is highly specious. Contemporary
scholarship has emphasized that, in Genesis 1:26–27, the first term
to appear is ṣelem with the preposition *bet* attached; the preposi-
tional prefix attached to dəmût, which follows, is *koph*. In Genesis 5:3, however, the word order and the attached prepositions are
reversed (the word dəmût appears first with the preposition *bet* attached; it is followed by the word ṣelem with the preposition *koph*
attached). The interchanged word order (with interchanging prepositions as well) suggests that syntactically these terms interpret each
other. Any contemporary effort trying to derive significant theological meaning for the *imago Dei* that is based on a hard exegetical distinction derived from the semantic value of these two terms should
be met with a great deal of skepticism.[4]

Nonetheless, it has *traditionally* been by exegetically separating the semantic value of these two terms that significant theological reflection on the *imago Dei* has proceeded, and, *theologically*, that distinction has been important. Irenaeus is the first of the early church thinkers to substantially address the issue of the *imago Dei* and propose just such a divide. His account of the doctrine appears in his work *Against Heresies,* wherein he is most explicitly concerned with refuting Valentinian Gnosticism and Marcionism. Irenaeus worried that both heresies adopted a view of creation and salvation that was not amenable to the biblical witness. In the case of Valentinian Gnosticism specifically, he found an unsupportable separation of the creative and salvific work of God.[5]

Even a cursory knowledge of Irenaeus and his account of recapitulation should make clear why he finds the separation of creation and redemption untenable. The Christ's salvific work fulfills the *promised* glory of creation manifest before the fall of Adam. It is promised because this glory of creation is only fully instantiated in the yet to be incarnate Christ. Beginning from this supposition, the Christ is the new Adam working as a soteriological presence within an unfolding order of creation that recapitulates and makes wholly manifest the originally promised glory of creation.[6]

The logic of Irenaeus's recapitulation entails a form of *logos* christology that in turn relies on a specific understanding of the *imitatio Christi.* The imitation at work in such a formulation is two-fold: employing the logic of how a reciprocal principle can harmonize two disparate concepts (God and human being) through a common element (the Christ). The sorts of imitation described by Irenaeus that allow the Christ to fulfill this mediating function make recourse to the doctrine of the *imago Dei.*

The logic works as follows. Despite the distance between the fleshly finitude of humanity and the infinite power of divine transcendence, the doctrine of the *imago Dei* indicates that these discordant terms must somehow be in harmony. The event of incarnation does the harmonizing. Insofar as the incarnation binds the *logos* to the flesh, Irenaeus suggests that God is made available to humanity. This binding occurs through a two-fold imitation enacted by the Christ. First, the Christ imitates Adam in fleshliness, thereby harmonizing God with human being. Second, the Christ simultaneously imitates God in obedience (justifying humanity before God), thereby harmonizing human being with God.[7] For Irenaeus, when the Genesis account refers to human beings as the image of God, it is the Christ—who is yet unrevealed—that is the proper object of this ascription; the salvific work of God done in Christ is a recapitulation of the well-ordered perfection of original creation where Christ is implicitly present as the *imago Dei*.[8]

Certainly, Irenaeus's account of the *imago Dei* and his understanding of recapitulation are strongly indebted to Pauline themes related to theological anthropology and christology. Specifically, Paul's use of the *imago Dei* serves as an imaginative projection of the idealized human being: The *imago* is the human being God transformatively calls each person to become in imitation of the Christ. Theological anthropology is thereby sieved through a specific account of the eschatological hope implied by christology; a doctrine that was a feature of creation gets displaced into the eschaton.[9]

What is distinctive about Irenaeus's approach is the way he ties these concepts to a formal, exegetical distinction between the "image" and the "likeness" of God in Genesis. He uses this distinction to indicate how the *imago Dei* is not *solely* relevant as an eschatological projection. The image of God consists of those qualities of similitude to the Christ that *persist despite the fall*, while the likeness of God refers to those qualities that might be *taken up as a fulfillment* of what was originally promised in creation but can be enacted only through subsequent imitation of the Christ after the fall. Now it is easy to follow how Irenaeus becomes more specific about what constitutes the *persistent image* that stands in distinction to the *likeness attained* in eschatological fulfillment. The image of God, for Irenaeus, refers to the powers of rationality and freedom expressed through the unity of human body and soul; the likeness of God denotes the presence of the divine spirit in redeemed humanity.[10]

The advantage of such an approach for Irenaeus is that it provides a tidy schema, seemingly revealed within scripture, which resists a gnostic denigration of the body and, by extension, creation (which the Pauline displacement—taken in isolation—could not explicitly rule out). Insofar as the Christ is incarnate and constitutes the perfect image of God we imitate, then there cannot be a separation of body and soul—wherein the material body is a simple hindrance to salvation.[11] The body also must be a part of this image of God insofar as the Christ has a body and imitates the fleshliness of Adam. Salvation is not a release of the soul from the body but a recapitulation of body and soul toward the divine *spirit*. In this way, the creative and salvific work of God remain tightly correlated.[12]

Despite running counter to the contemporary suspicion for proposing significant semantic distinctions between "image" and "likeness" in interpreting the *imago Dei*, there is both a historical and an existential importance to Irenaeus's schema. Historically, Irenaeus elucidates contours that structure subsequent theological development of this doctrine. His clear exegetical distinction between "image" and "likeness" is taken up repeatedly in other early church sources. While his precise formulation of how one constitutes each of the terms at stake in this distinction will be changed by various thinkers, the fundamental sensibility that the image indicates an indissoluble connection with the divine that can then somehow be augmented by recovering a lost sense of likeness remains a foundational concept. There is not a significant alternative approach to Irenaeus's exegetical distinction until Augustine.[13]

While repudiating the untenable exegetical distinction between "image" and "likeness" has become a modern trope for how to characterize Irenaeus's account of the *imago Dei*, we should give more credit to how his account represents an act of theological imagination that responds (in the conceptual language of his own context) to a significant, existential question: *What does it mean that we are not yet what we are supposed to be?* If we understand Irenaeus's account as a response to this question, we could (anachronistically) say that he is a proto-existential thinker in the broad sense that Paul Tillich uses the term: one who is concerned with the immediate personal experience of existence—*Innerlichkeit*.[14]

Constructive theological efforts dealing with the *imago Dei* must take biblical scholarship seriously and abandon pursuing a formal exegetical distinction between ṣelem and dəmût. However, this does not mean we should abandon the theological force of Irenaeus's insight, which speaks to a persistent sense in our own experience: We are not yet what we ought to be. In dealing with this question, Irenaeus's distinction between "image" and "likeness" lays out lines that help demarcate the field of play for doing constructive work on the *imago Dei*.

On the one hand, whatever it means to be the image of God cannot entail so much perfection that we lose sight of sin; we may be the image of God, but we are far from perfect. To be the image of God is not an attestation of a presently accessible unmarred human perfection. On the other hand, the image cannot be totally projected into the future as an eschatological end disconnected from our present reality or utterly lost in a metaphorical interpretation of the Fall. An account of the *imago Dei* that falls either to perfection or depravity is insufficient. This is the theological force of accounts of the *imago Dei* that make a distinction between image and likeness, no matter how exegetically untenable: They help lay out the boundaries of reasonable thought about the *imago Dei*. A constructive effort at rendering the meaning of this doctrinal symbol, one taking stock of the immediate personal experience of existence characteristic of our own time, must still produce concepts that are flexible enough to accommodate the sense that we are not yet what we are supposed to be, but not so far from what we are supposed to be as to exist in discontinuity from our hope.

## Image as Symbol

When introducing the *imago Dei* to a class of undergraduates, I was surprised when, halfway through the period, a bright student, eagerly describing himself as spiritual but not religious, suddenly spoke up and said, "Oh, this is about human beings!" Because I thought it was clear that the topic for the day was theological anthropology, I was a little underwhelmed by what he took to be a startling revelation. Pointing this out to him with a look of perplexed irritation, he responded, "I knew the topic for the day; that's why I thought it was weird that you have been talking about what people think God looks like. You know, the *image* of God." To my own amazement, a surprising number of his classmates nodded in agreement.

We all too easily forget that someone encountering this doctrine without sufficient awareness of its biblical context may hear the word "image" in a much more quotidian sense. "Image" generally conveys some sort of imitation or representation composed in a different, artificial medium. The word brings to mind a photograph, sculpture, or painting—a picture of some kind—perhaps even more specifically, it makes us think of portraiture. If I told my children, "Here is an image of your grandmother," and I read a story about her buying clothes for them, they might make a connection that the story was meant to demonstrate the love and care she feels for them, but this certainly would not be the first thing they expected when told they were being given an image. Our quotidian assumption is that an image is like a mirror; it reflects back a likeness to the thing being imaged. In this sense, image implies some sort of fundamental similarity to that which is being imaged that we can recognize. An image "figures" with some degree of semblance.

Yet, if we look at the use of the word "image"—ṣelem—outside the primeval history of Genesis, we get a sense that semantically this term indicates something richer than a mere mirroring likeness or reflective semblance. Generally, it does refer to some kind of formative replica. This might take the shape of an image, idol, or picture (or even more figuratively a dream).[15] Prototypical examples might include Numbers 33:52, 2 Kings 11:18, 2 Chronicles 23:17, and Amos 5:36, where the term refers to an idol of another God. However, the oldest preserved use of the lexeme in the Hebrew Bible probably comes from the ark narrative in 1 Samuel 6:5, 11; this instance serves as a good reminder of the magical quality this term connotes.

In this passage, the images of boils (or hemorrhoids) and mice cast in gold are symbolically "driven away" by the Philistine rulers who had been cursed while in possession of the ark. The hope is that by sending out the ark with the images of the boils and mice as a compensatory offering, the actual things cursing the Philistine cities themselves might depart as well. The "images" employed in this passage from 1 Samuel hold a much deeper meaning than what we typically attribute to the term: the "image" of this passage implies "that which is depicted is itself present."[16] The mice and boil images are meant to take on the actual pestilent aspects of life in the community and purge it for the sake of well-being.

When looking to these instances of the wider use of the term ṣelem, we find it has a semantic range closer to the way we might use "symbol" as a technical term in theological discourse: the symbol, as opposed to the sign, participates in that to which it points. The participation opens up a level of interconnected reality (in the case from 1 Samuel a control over the curse) that is otherwise unrealized. In this sense, the image (ṣelem) as symbol indicates an indelible connection that is deeper than mere reflection; actions taken on the image affect whatever the image points to.

To further this understanding, we can consider the use of *ṣalm (as the Proto-Semitic root) in other ancient Near Eastern cultures. In both the Akkadian realm and the eighteenth dynasty of the Egyptian pharaohs, a word from the root *ṣalm describes the king's relation to the deity. In the case of the Egyptian pharaohs, the ruling monarch becomes the figurative inception of the divine (being the image of Bel or Ra respectively); thus, the territory is ruled over by the monarch as an acting proxy for the divine, who thereby enacts the power and will of the deity. In a visceral way, the king or queen acts within his or her commissioned realm as a symbol participating in the divine: drawing on the full authority of the deity to whom he or she points and thereby acting to extend the power of the referenced deity itself.[17]

Similarly, the statue of King Adad-iti found in Syria in 1979 can be helpful in interpreting the meaning of ṣelem. It was common practice amidst Near Eastern kings to create images of themselves as symbols of their rule in conquered lands from which they were absent. The statue of King Adad-iti is particularly interesting because it has an Akkadian and Aramaic inscription. In the Aramaic inscription, the equivalent cognates to both ṣelem and dəmût appear, while the Akkadian inscription has the equivalent cognate to ṣelem. The inscription on the statue indicates that it was formed to honor the God Adad in thanksgiving for Adad-iti's rule over Guzan. Further, the statue is intended to "perpetuate the throne" of Adad-iti. This particular statue gives us a glimpse of the chain of symbolic power wherein the statue is a proxy for Adad-iti, and Adad-iti is a proxy for Adad. Moreover, the curse in the Aramaic inscription gives a sense of the severity of punishment for defiling the statue. This severity of the curse is logical when one understands the statute to be a visceral proxy that *participates* in the power of the king over the kingdom and that the king is a visceral proxy that *participates* in the power of the God over the people. To defile the statue is to defile the power of the king and, by extension, the power of the God.[18]

My aim is not to make a linguistic case for why ṣelem should be understood as symbol instead of image. I want to illustrate only that the semantic range of the term provides flexibility for the theological imagination. Image can be a remarkably participatory concept, certainly much more participatory than what is implied by the quotidian meaning of image. If our aim is to develop the meaning of this doctrine itself constructively as a symbol that aids people in ordering their existence toward ultimacy, then we need to start thinking beyond an understanding of the *imago* as merely making simple recourse to mirroring or reflection. How is it theologically efficacious to understand humankind as a "symbol of God"—human being itself participating in and pointing toward the divine? How are we an *imago* or a ṣelem that figures the divine not by appealing directly to some degree of semblance (the quotidian sense of an image), but one that figures the divine as an extension of God's features and powers?

Aspects of this idea are not utterly at odds with the theological development of this doctrine. Features of the *imago Dei* taken as symbolic participation appear in the work of thinkers such as Augustine and Friedrich Schleiermacher. Each points to features of human ways of being in the world that affect a participation in or with the divine. In the case of Augustine, such participation is a form of direct mirroring of God's own acts; however, in the case of Schleiermacher's God-consciousness, we get a glimpse of participation as being structural instead of merely reflective.

## Augustine

Perhaps no figure has had more influence on the doctrine of the *imago Dei* than Augustine. He does not make use of the formal exegetical distinction between "image" and "likeness" that was employed by so many thinkers after Irenaeus. Certainly, the image of God in human beings is not indicative of our perfection and he makes clear that human being is restored in relationship to the divine through the Christ; for anyone with a passing familiarity with Augustine, this is no surprise. Augustine's account of the *imago Dei* is distinctive, though, for how he imagines human beings imitating the trinitarian structure of God.[19]

This idea is the focus of the second half of Augustine's *The Trinity*. Here, he offers a rational explication of the trinity as it accords with various trinities found in the human mind.[20] Still, Augustine is careful to remind us that this speculative endeavor is a tenuous affair. There is no way to judge the full affinity of these mental trinities to the Godhead itself; there is no deductive path of natural theology running directly from the *imago Dei* to a trinitarian vision of God.[21] This does not indicate, by any means, that elaborating these psychological trinities is a frivolous endeavor, unnecessary in light of the occurrence of the trinity in scripture.[22] The trinitarian ordering of human being as *imago Dei* makes aspects of the trinitarian structure of the divine available to us experientially and analogically; it makes the trinitarian structure available to us in ways that go beyond its formal revelation in scripture. Moreover, if our trinitarian knowledge of God was limited solely to scriptural appeal, such knowledge might simply be nonsensical. This analogous and experiential knowledge of the trinity given through the structuring of the *imago Dei* is quite crucial.[23]

There is an important connection to be introduced here between the description of image as symbol offered previously and this availability of the trinitarian structure. If a symbol functions to convey a means of ordering existence that makes ultimacy available to our participation (via a given context of engagement through which the symbol is interpreted), for Augustine something similar seems to be at stake with the trinitarian structure of the *imago Dei*. The structure of our mental faculties as they encounter the world provide an analogical and experiential participation in God's trinitarian life.

This symbolic encounter of the divine trinity through the *imago Dei* is deepened if we note that Augustine employs a common distinction in predication to describe both phenomena: a distinction between substantial and relative predication. He used the distinction to argue against Arianism, and it is a distinction that is simple to rehearse.

First, we have to assume the common axiom of his time that since there is nothing changeable in God, nothing can be said of God according to accidents or modification. The Arians took this to mean everything predicated of God was claimed in terms of substance, the consequence being that all claims of difference between God and the Christ were indicative of real differences in substance (notably that because Christ is begotten of God and God is unbegotten, Christ represents a lower substance). Augustine counters this mode of argument by claiming there are two modes of proper predication about God: substantial and relative. Relative predication is not concerned with accidents (thereby preserving the unchangeable quality of God), but refers to *unchangeable relationships* within the Godhead: "Therefore, although being Father is different from being Son, there is no difference of substance, because they are not called these things substance-wise but relationship-wise; and yet this relationship is not a modification, because it is not changeable."[24] All other predications made of God that are not substantial or relative must be understood metaphorically.[25] This understanding of relative predication also sets the ground rules for describing those trinities that we find in humankind if we are the *imago Trinitate*, giving a more specific sense to what our participation in the divine as a symbol of God might entail on Augustine's view. Simply put, the *imago Dei* indicates where relative predication is applicable in the life of a human being.

If proper predication about the divine is made in terms of substance and relation while all other speech is metaphorical, any *acci-*

*dental* references in the trinities of the *imago Dei* would not be parallel to the trinity proper. Reference to that which is accidental could only be parallel to metaphorical speech, not to issues involving substantial and relative predication. As such, Augustine rejects any *corporeal* representation of the relationships of the trinitarian *imago Dei*. It is impossible to eliminate accidental references with bodily relations and thus maintain a proper analogy to the relative predications of the divine trinity.[26]

To find a proper trinitarian analogy in the *imago Dei*, Augustine begins with scripture (as God's own self-revelation). A substantive predication about God from scripture ought to yield a set of corresponding relative predications that could be analogically manifest by human being if we are truly the image of the Trinity.[27] Augustine's substantial predication affirms that God is love itself (referencing 1 John 4:7–21).[28] If God is love, then when we act out love Augustine suggests we are in God who is love itself. In loving, we are the *imago Dei* by analogy to and experience of that which is substantially predicated about God.

This can be pressed further, however, by looking introspectively at this act of love. Looking introspectively at love, we also can come analogously and experientially to know God's trinitarian life. In short, our act of loving (by which we are the image for what can be substantially predicated of God) works according to a three-fold set of relative predications that Augustine takes to be the way that we are the image for the relative predication about God's trinitarian life.[29] The three relations in love are as follows: "Now love means someone loving and something loved with love. There you are with three, the lover, what is being loved, and love."[30] Introspective examination of love reveals the *imago Dei* in acts of love based on the presence of an analogy that can be drawn to both the substantive *and* relative predications concerning the divine. We are the *imago Dei* as an *imago Trinitate* in loving: The love itself is an image of what can be substantially predicated of God and the three-fold relationship at work in all loving is an image of what can be relationally predicated of God.

Augustine will extend this idea much further, but we need not follow him in delving more deeply into the immutable relations of the mind, knowledge of the mind, and love of the mind.[31] Our concern is more simply for the logic of Augustine's claims regarding how it is possible for one to formulate the doctrine of the *imago Dei*. What we find is a clear argument for understanding what it means to be the image of God based on analogy to statements made about the divine life. The analogy affects a certain sense of symbolic participation in/with the divine; our existence as the *imago Dei* opens up a means of experiencing the relative predication at work in an account of the trinity that would not otherwise be available in all of creation. Being the *imago Dei*, humankind is the analogical symbol by which the substantial and relative predication at work as a feature of the divine life is recognized and then expressed in creation.

## Schleiermacher

The constructive facets of Friederich Schleiermacher's understanding of the *imago Dei* occur in a few tightly reasoned pages. The doctrine's placement in relation to other loci of systematic theology, in his seminal volume *The Christian Faith*, is both crucial and distinctive. The placement reflects Schleiermacher's context: writing in the wake of the Enlightenment's rationalism, which had stripped bare metaphysical presuppositions of natural theology and questioned the historical and moral reliability of Scripture as the authoritative foundation for doctrinal theology.[32] In such a context, theology can no longer arrange doctrinal statements by simple appeal to a speculative metaphysical foundation, nor can scripture serve as a settled history of unquestioned authority for such arrangement.[33]

It is out of these eighteenth-century developments that we must understand Schleiermacher's work: The theological object must be approached with absolute diligence and serious consideration given to the experiencing religious subject. As such, we find in Schleiermacher a fundamental reorientation of the theological paradigm to "feeling" or "immediate self-consciousness" as the essential seat of all piety. This interpretation of feeling is coextensive to all of our willing or thinking as an essential quality in it. Feeling is *not* a faculty parallel to our willing and thinking.[34] Instead, feeling indicates the affective immediacy of existent being apart from objective contemplation.[35]

Because Christian theology after the Enlightenment cannot be pursued in its systematic ordering along self-evident or historical principles, it instead must be accounted for in terms of possible effects on affective, immediate self-consciousness. To do this, Schleiermacher employs two organizational schemas, each of which are three-fold.[36] According to the first schema, the whole range of Christian doctrine is addressed in articulating the feeling of religious self-consciousness as it is factually expressed by the central Christian antithesis between (1) sin and (2) grace, if sufficient attention is also devoted to (3) those facts of religious self-consciousness that are unchanged but presupposed by this Christian antithesis.[37] According to the second schema, grace, Schleiermacher suggests all the doctrines in Christian systematic theology (understood as implications of the feeling of absolute dependence characteristic of affective, religious self-consciousness) are directed toward three leading ideas: (A) to human states of life, (B) to divine attributes and actions, or (C) to the constitution of the world. Schleiermacher characterizes various doctrines typically dealt with in Christian theology in terms of their intersection with each of these three-fold schemas. One could even sketch this out as a simple Punnett square.

Looking at the place of the *imago Dei* in relation to these schemas helps illustrate why Schleiermacher's theological approach can feel a bit foreign at first. For instance, many systematic theologies from before the Enlightenment were organized according to the arc of salvation history. If Schleiermacher were to simply follow this traditional approach (which is specifically what he says *cannot* happen given his context) but using the terms of his two-part, three-fold schema outlined above, we would expect that the *imago Dei* would be addressed as part of the account of sin (as the preliminary state being changed) and directed toward dealing with states of human life (1A). *This is not what Schleiermacher does*. Instead, he suggests the *imago Dei* is an unchanged but presupposed fact of religious self-consciousness directed to the constitution of the world (3C). Charting this out on a Punnett square, the treatment of this doctrine is as far as it possibly could be from where we would have expected to find it in previous systematic treatments.

It is this second part—that the *imago Dei* is directed toward the constitution of the world and not a human state of life—which is most surprising. Schleiermacher associates the doctrine of the *imago Dei* with a state of originary human perfection, wherein humankind and the world belong *inseparably together*. Importantly, though, the "perfection" Schleiermacher is imagining is the necessity of any thinking or willing experience to yield religious affection. The perfection is one whereby all immediate self-consciousness is concurrent to God-consciousness.[38]

In any experience of immediate self-consciousness (that seat of piety around which systematic theology should be ordered) there are a pair of common elements: one self-caused and one not self-caused—a Being and a Somehow-having-become—*ein Sein und ein Irgendwiegewordensein*.[39] These two elements correspond to the activity and receptivity of the self-consciousness. All of those expressions of the receptivity of the self-consciousness—our coexistence with an Other—have a common, essential feeling, the feeling of dependence; this feeling is absolute in that it accompanies the whole of our self-conscious existence.[40] This feeling of absolute dependence is the same thing for Schleiermacher as being in relation to God, wherein God is the "whence" co-determinate to this feeling of dependence given over to us in all aspects of our immediate self-consciousness. In this way, to experience the feeling of absolute dependence is God-consciousness.[41]

If the *imago Dei* describes a state in which all immediate self-consciousness is concurrent to God-consciousness, then what is at stake is an attunement between human being and the world that characterizes a ground on which the antithesis of sin and grace can be thematized. The perfection of the world refers to the necessity of world-impressions to be of such a quality that God-consciousness can be excited by them; the perfection of humankind refers to the constitution of our self-consciousness such that these impressions excite religious affection specifically. One without the other is insufficient; both facets of this perfection must exist for either to be meaningful.

Moreover, it is not enough that immediate self-consciousness be able to excite religious affection just once or twice. The realization of God-consciousness must be *continuously* possible if we exhibit an original perfection of attunement between world-impressions and God-consciousness. The *imago Dei* must be a "living impulse."[42] To affirm anything less than this continuous possibility would make for a dissonance incongruous to the perfection the *imago Dei* implies. If the attunement were discontinuous, it would mean the feeling of absolute dependence—the root of God-consciousness that is a stalwart feature of all immediate self-consciousness—arises only across scattered moments of our lives; it would entail a fissure in the very experience of immediate self-consciousness: a divide between moments where God-consciousness is possible and those where it is not.

Perfect God-consciousness would occur when our immediate self-consciousness continuously excited religious affection *in actuality*. The claim of the *imago Dei* is more modest: An immediate self-consciousness wherein world-impressions and God-consciousness are attuned is continuously *possible*.[43] Yet even this possibility indicates a human distinctiveness in comparison to the rest of creation (historically this is the essential feature Schleiermacher suggests an account of the *imago Dei* ought to elucidate).[44] Being the image of God entails the very possibility of participation in the world such that God-consciousness is available.

For all the drastic differences between the accounts of the *imago Dei* found in Augustine and Schleiermacher respectively, in each we find features of an understanding of the *imago* as symbol. While Augustine's approach is introspective and Schleiermacher looks to the border between self and world, each is thematizing the *imago Dei* in such a way that humankind (either through the trinity or God-consciousness) becomes a symbol through which creation continues to participate in the divine life and makes the divine life available in the midst of creation in distinctive ways. For Augustine, in loving we bring an experiential and analogical knowledge of relative predication at work in the divine life into the world. For Schleiermacher, our existence makes an awareness of God's presence to creation continuously possible through our immediate self-consciousness. In both cases, to be the image of God invokes a sense of participation (as with a symbol) in the divine that reveals something of this ultimacy that might otherwise remain ineffable.

## Refracting the Divine

How do we begin bringing these various threads together? Using astrobiology as the context of engagement for interpreting the significance of the *imago Dei* can provide an initial way out of the rut of human distinctiveness that too easily overwhelms the interpretation of this doctrine. Even before looking to the history of the doctrine's interpretation itself, acknowledging the centrality of intra-action between living-systems and habitability puts a check on efforts to categorize human being in terms of some stipulative distinctiveness. Astrobiologically, a living-system is not its own; its distinctiveness is always given over as part of the wider intra-active phenomenon that includes its habitable environment. Whatever property might be used to characterize a given living-system should be able to be reframed in ways that give primacy to the distinctiveness of the habitable environment if the mutuality of intra-action is as thoroughgoing as was suggested in the previous chapter.

However, looking back to historical interpretations of the doctrine, particularly looking to how the doctrine intersected with other doctrines such as creation, christology, and trinity to name a few, we can begin to see an alternative story for interpreting the *imago Dei* that can be coherent with the astrobiological context of engagement. To conclude, I want to lay out the crucial steps or ideas that emerge from recontextualizing these historical interpretations in light of astrobiology.

We can begin with Irenaeus. His account reminds us that the *imago Dei* has to speak to the angst of our immediate personal experience of existence—*Innerlichkeit*—in which we wrestle with the suspicion that we are not yet what we are supposed to be. A constructive account of the *imago Dei* needs to begin by taking account of this existential tension that was an axiomatic feature of his interpretation of this doctrine.

Irenaeus's clear exegetical distinction between "image" and "likeness" was meant to address this tension while simultaneously speaking to concerns about the status of the material body in the Valentinian Gnosticism of his day. The image is not something lost to humankind in the utter depravity of the Fall; we continue—in the created goodness of a unity between our material bodies and rational souls—to bear the image of God. The likeness is a recapitulation of what is first imitated for us in the Christ and promised in creation but never realized on account of the Fall: the full presence of the divine spirit.

Certainly, the exegetical distinction between "image" and "likeness" is no longer tenable. Moreover, Irenaeus's approach to the *imago Dei* was constrained by a certain prelapsarian versus postlapsarian logic that took the Fall as a historical and axiomatic point of transition for human being in a wider narrative of salvation history. But the existential reality that we are not yet what we are supposed to be—that the *imago Dei* is something we may already be, but also something that in its fullness must be lived into—remains a crucial ground from which to develop a constructive account of this doctrine.

We might next take a cue from Schleiermacher and emphasize the need to divorce our understanding of the *imago Dei* from a highly specific account of salvation history. His account still emphasizes the need for our formulation of the *imago Dei* to take seriously the tensions of our immediate experience of existence because we experience ourselves as not being what we ought to be insofar as every world-impression is only *potentially* but not *actually* attuned to God-consciousness. With Schleiermacher, the *imago Dei* is imagined as a structural condition of our way of being in the world. This reshaping divorces the *imago Dei* from an axiomatic reliance on originary perfection that ties its meaning to a specific narrative of salvation history.

To put this in terms that might be useful going forward, we need to thematize the *imago Dei* as something that, simultaneously, we *both are and strive to become*. It is a condition that may be more or less realized at given times, a condition to which we are more attuned or from which we are more estranged. We are never utterly divorced from bearing the image of God, but it is also something we must continuously work toward realizing. As a first step toward a constructive account, we should address both how we are the *imago Dei* and how we work to become the *imago Dei* as distinct but non-separable ideas that emphasize how this doctrine is a structural principle applicable to the intersection of human being with the world.

To get more specific about how the *imago Dei* is something that is, grammatically speaking, both present perfect (something having been the case in the past and continuing into the present) and future (something we are not yet but hope to become), we can turn to the "*Dei*" in this doctrine. Thinking back to the trajectory of the cosmogonic tradition that orients the *imago Dei*, as described in the previous chapter, God—that which is ultimate—imputes creation with the possibility of having a meaningful existence. Cosmogony does not chronicle the production of sheer existence; creation is a story in which determinate, particular creatures carefully intended for *meaningful* existence come to be.

Further, I emphasized that the world of cosmogonies is one in which this pursuit of meaningful existence is by no means a Sisyphean task. There is a persistent ordering to creation wherein God attentively provides for the needs of creation so that this possibility of meaningful existence may be actualized by creatures. The order is never so encompassing as to smother the freedom of creation. Nonetheless, this God is one whose very disposition is oriented toward transforming chaos into the consistent order of the cosmos so that *well*-being, not just some banal material being, is possible.

Thus, the previous chapter argued for a link between cosmogony and astrobiology. Cosmogony reveals a vision wherein God's creative power exemplifies an aesthetic proclivity for the sorts of intra-action between living being and its habitable environment that is the subject of astrobiology. In both cases, it is the quality of this intra-action that makes the flourishing and vibrancy of meaningful existence possible. To put a finer point on this claim, I am suggesting the meaningful existence of well-being we find in cosmogonies is a narrative expression of the intra-action that astrobiology studies as a potentially emergent feature characteristic of all living-systems in the universe. For both cosmogonies and astrobiology, living things do not merely exist; they live toward a specific sense of well-being that is irreparably connected to the wider habitable environment.

In the context of the *imago Dei* this becomes significant in two senses. First, human being is located squarely within this cosmogonic picture. We are a creature, in the language of cosmogony, or a living-system, in the language of astrobiology, engaged in the intra-active pursuit of meaningful existence. Second, as a creature that is the *imago Dei*, it is this God—the God who is not a being but being-itself, the divine that is an expression of ultimacy through the *prius* of creative power working toward establishing fruitful order across the whole panoply and plethora of existence, the God that with a transformative disposition gives space for creation to freely develop this meaningful existence made possible in creation—of which human being is the image. The creative power that is exhibited by God in these mythic stories provides a frame to the understanding of God in which we are the image; God is the indeterminate, transcendent power of being-itself that is bringing forth the possibility of meaningful existence and supporting the free formation of such meaning through a careful ordering of creation.[45]

Herein, to be the *imago Dei* is not to be concerned with thematizing some sense of distinctiveness from the rest of creation, but to better understand our relationship as human beings to the creative power of the divine that has been, is, and will be—a creative power that is vested in providing well-ordered relations between creatures and environments that makes establishing meaningful existence possible. So a second step in offering a constructive account of the *imago Dei* relies on making reference to the creative power of the divine as that of which we bear an image. Reference to this creative power provides a continuity between how we simultaneously are and become the *imago Dei*, and it is coherent with the intra-action characteristic of astrobiology.

Now, we can turn to the "*imago*" itself. In its quotidian sense, image conveys a semblance figured in a different medium; it is a mirroring likeness (even if imperfect) where to be an image is to appear like that which is being imaged. However, what we found was that image can also entail more than this rudimentary "appearing like" that we too easily ascribe to the *imago Dei*.

Whether figured in terms of statues extending the power of a king, in terms of introspectively analyzing love as a trinity that analogically extends predication about the divine life into our experience of the world, or in terms of the continuous possibility by which immediate self-consciousness is God-consciousness, there is a strain of thinking in the theological tradition for interpreting the *imago Dei* that emphasizes its meaning in terms of how human beings are able to extend powers and features of the divine into the world. To be the image of God is to affect the kind of participation characteristic of a symbol: Human being makes the ultimacy characterized by the transcendence of divine creative power an immanent concern available in the midst of creation.

It is important to be clear about this point because the analogies of mirroring or symbolizing imply very different things about the way we engage the creative power of the divine. If we claim that to be the *imago Dei* is to mirror the divine (no matter how darkly), the analogy drawn is to a specific kind of reflection—a concept that can be well understood in terms of some very simple physics. In specular reflection (such as when you see yourself in a mirror) a light wave (the incident wave) hits a barrier and bounces off it (now called the reflected wave); the law of specular reflection states that the angle at which the incident wave approaches the barrier will always be the same as the single angle at which the reflected wave rebounds.[46]

When we consider the *imago Dei* with this more robust understanding of how reflection works, there is something dissatisfying about such an analogy. The light waves in specular reflection simply bounce off the mirror as a boundary or barrier that is not passed. As the image of God, are we such a barrier? Are we a place in creation where the creative power of the divine simply bounces off us and moves in new directions? If "image" connotes symbol, we should expect something more participatory. We engage with this divine creative power.

As an alternative optic analogy for this deeper engagement, we might look to a different way light behaves when it reaches a boundary: refraction. In refraction, waves encountering a barrier do not bounce off it; instead, the wave passes through the barrier from one medium to another. As the wave passes through the barrier into the new medium, its speed may change. The change in speed bends the wave. Such bending and changing can have surprising effects; for instance, it is refraction that is at work when light hitting a prism scatters rainbows.

How a prism disperses light by refraction is relatively simple to rehearse. Visible light, or white light, consists of a collection of component colors across the spectrum. Each color is characterized by a distinct wave frequency. Dispersion occurs because the distinct wave frequencies of the component colors will bend by varying amounts as the white light slows down as it passes from moving through the air to moving through the prism. Technically put, the optical density of the prism is different than that of the air through which the white light was moving; this optical density is characterized in terms of the index of refraction value. That index of refraction value varies with light frequency. Thus, as the white light travels more slowly through the glass of the prism the various wavelengths of the component colors are separated because the angle of refraction is different for various wavelengths.

We can claim that we are like a prism not a mirror. *To be the image of God is not to reflect the divine but to refract the divine.* The *imago Dei* is not a boundary that the creative power of the divine bounces off. Instead, it is a medium through which the work of the continuing creative power of the divine passes through and is bent; and, in being bent seemingly invisible features of that creative power are dispersed and available in new ways.

We need to stop thinking of what it means to be the *imago Dei* in terms of a formative replica or a mirror image: an act of directly imitating God's creative power. Instead, to be the image of God is to exist as that place in creation that can disperse the creative power of God, in all its possibilizing of meaningful existence, in new ways. As Phillip Melanchthon artfully puts it in his commentary "On the Soul," "Thus I call the image of God the powers of the soul when God shines in them."[47]

The implications of this constructive account of the *imago Dei*, particularly shifting from reflection to refraction, are significant. Rather than simply asserting that human beings, in contrast to the rest of creation, are the *imago Dei*, now we can suggest that *any* creature that refracts the creative power of the divine—that extends the creative possibility for creatures to pursue meaningful existence in ways that would otherwise remain invisible or impossible—is the *imago Dei*.

This point is worth emphasizing. A refractive account of the *imago Dei* is not anthropocentric: The primary purpose of the doctrine is not to establish what constitutes our human uniqueness or distinctiveness in relationship to the other creatures we find here on Earth. So often, theological reflection about the *imago Dei* has focused on this point—uniqueness or distinctiveness—to the exclusion of other features of the doctrine. That nearly exclusive focus makes it tempting simply to cast this doctrine aside as untenable or ecologically harmful because it must be inherently anthropocentric. However, there are other ways to interpret the meaning of this symbol.

Further, it may seem so banal as not to need to be mentioned, but the refractive account of the *imago Dei* entails being a creature. That which refracts is a medium for that which is refracted; they are fundamentally different sorts of things. In a refractive account of the *imago Dei* there is a clear distinction maintained between the creatureliness of that which bears the image and the divine creative power itself. Moreover, not just any creature is the *imago Dei*. To bear the image of God means having one's own meaningful existence tied to this way of extending and making the possibility of meaningful existence present to creation in new ways: It is to be a creature that plays a critical part both in understanding and fostering a careful stewardship of intra-action between living-systems and their habitable environments.

Astrobiologically, interpreting the doctrine this way is important. Such a rendering leaves open the possibility that we could find other instances of the *imago Dei* either here on Earth or on other worlds.[48] It also leaves open the possibility that, as a species, we might continue to be the *imago Dei* as we evolve into some posthumanist future because the doctrine might apply to any set of creatures whose meaningful existence depends on facilitating the intra-actions of other living-systems with the wider habitable environment. Thus, a post-human species, a living-system intra-acting with a wholly different habitable environment, or a newly evolved species on this planet could all equally well understand their own meaningful existence in terms of refracting the creative power of the divine. If we interpret the *imago Dei* astrobiologically, it does not categorize human uniqueness or distinctiveness, per se; instead, it categorizes the intra-action pursued between a specific sort of living-system and its habitable environment. What exactly constitutes the image of God could be multiply realized and in principle plural in terms of its expressions (i.e., it may not look the same in all instances).

Finally, to act as the *imago Dei* entails a responsibility. We can fail in extending the possibility for meaningful existence. It is something that we are, as this is a fundamental part constituting our own sense of meaningful existence, and it is something we have to become, insofar as we actually do or do not refract the divine to open new possibilities for meaningful existence across creation. The same would be true for any other species we might describe as being the *imago Dei* anywhere else in the universe.

This is by no means an easy responsibility to bear or conceptualize. It entails a responsibility to reveal and foster the deeply interconnected network of intra-actions that would exist in any habitable environment with living-systems. If, as was stated at the beginning of this chapter, symbols connect us to an inter-human depth and condition the way we behave toward vulnerable, planetary others, the refractive account of the *imago Dei* can be an overarching theological symbol for this responsibility. It summons us to a particular way of being in the world that resists de-humanizing and de-valuing all sorts of planetary others in our responsibility to bend the creative power of the divine to facilitate the flourishing of astrobiologically significant intra-actions.

What has yet to be considered, however, is the scale at which being the *imago Dei* applies to humankind. In this chapter, we have focused on how the intra-action at work in astrobiology can be the context of engagement for this symbol, reclaiming aspects of what the symbol might mean when its interpretation is freed from an overwhelming concern for delineating human distinctiveness from the rest of creation. Yet, in the context of astrobiology, we must also emphasize that it is not our organismic particularity that is of primary interest but an understanding of our species as a living-system.

This means that an effective account of the *imago Dei* cannot just look inward toward the communities from which this doctrinal symbol originally arose. Living into the *imago Dei* is not something that can be achieved by any single religious individual, community, or even tradition. To be the *imago Dei* as humankind means thinking carefully about how we must be open to working with all sorts of other persons who might orient around religion in quite different ways, but share in the responsibility and human accountability that are part of the calling of the *imago Dei* toward facilitating the flourishing of intra-actions between living-systems and the wider habitable environment.

So what happens when we think of the *imago Dei* as applying not to individuals or small groups, but the whole human species? Can we even conceive of the entirety of the species refracting the creative power of the divine? How would we characterize agency and responsibility in such a situation? In a world of climate change increasingly characterized by language about the Anthropocene, these sorts of questions have become ever more pressing.

# CHAPTER
# 5

# Conceptualizing Nature

The previous chapter began developing how the *imago Dei* is a symbol, with astrobiology forming the context of engagement. While it seemed reasonable to assume that considerations of the intersection of the *imago Dei* with astrobiology would proceed in parallel to existing inquiries in theology and science, which examine the intersection of this doctrine with other scientific fields, this proved problematic. Typically, being the image of God connoted a human uniqueness in distinction to the rest of creation that was thematized in terms of some substantial, relational, or functional characteristic of an individual person. The prevailing logic was that to be the *imago Dei* meant an individual mirrored some aspect of the divine that stipulated our distinctiveness from the rest of creation.

Astrobiology's concern for the intra-action of living-systems and the habitable environment provides a context of engagement for interpreting the *imago Dei* that is significantly different. Reimagining how elements from previous interpretations of the *imago Dei* could be consistent with astrobiology, I suggested that to be the *imago Dei* was not to *reflect* the divine, as though we mirror some divine quality, but to *refract* the divine, as though we were a lens through which the creative power of God passed. Striving to live into a well-ordered relationship with creation, we become not simply co-creators that mirror divine power, but agents that channel the continuing creative power of the divine extending the possibilities for the meaningful existence of creation: revealing a way of God's being-with creation in its flourishing that may have otherwise remained hidden. Dispersed through us, this creative power of the divine can be used in new ways—it makes new ways of conceiving of meaningful existence possible—and it indicates new sorts of responsibilities for the *imago Dei* that is refracting this power.

Looking back to the various historical treatments of the *imago Dei*, we found these treatments assumed the image of God operated as a property of individuals. To think of this doctrine in terms of the astrobiological context of engagement, we must get beyond this fetishization of the individual as *imago Dei*. It would not be you or I as individuals who are the image of God, but human being understood as a living-system attuned to its habitable environment that can refract this image.

How, then, do we begin to think about human being broadly, not human individuals, as the lens that refracts the continuing creative power of the divine? What does it mean for individuals to participate in being the image of God while not individually being this image themselves? What is the scale of problems the doctrine of the *imago Dei* addresses if it is not localized to the individual?

The next two chapters begin dealing with these sorts of questions by turning to the concept of the Anthropocene as it can be interpreted with astrobiology as its context of engagement and in corroboration with the interpretation of the refractive *imago Dei* developed in the previous chapter. For nearly two decades, the term "Anthropocene" has been appearing with increasing frequency; what exactly it means is a bit difficult to pin down. The most specific sense invokes "Anthropocene" as a stratigraphic term that indicates various human-driven alterations to the material world are having a global effect evident in geological history—it is a technical concept within geology.[1] The effect is so significant geologists are deciding if we have transitioned out of the previous epoch (the Holocene). The Working Group on the Anthropocene (WGA) is developing a recommendation to the International Geological Congress that makes this case: setting apart the Anthropocene as a new geological epoch that began around 1950.

The recommendation of the WGA is still in its early phases, but it must do much more than offer up an approximate date at which the proposed geological epoch begins. The group has to present evidence there is a discernible and global signal within sediment layers or deposits that will remain in the geological record in order for this shift in epochs to be officially adopted. Radioactive elements from nuclear testing, carbon spheres from power stations, and plastic pollutants are all possible signals being considered by the WGA.[2]

While any official change remains a work in progress, there is undoubtedly power in a name and in the act of naming. To name a new geological epoch has wider repercussions that the popularization of the term "Anthropocene" is building upon. It indicates that human activity on the Earth is so pervasive that it rivals the greatest forces of nature that we know. Popularly, it has come to operate as a slogan for climate change. In any case, it definitively indicates what environmental writers have contended for a long time: that a straightforward division between human being and nature cannot be maintained.[3] In this sense, the term is not only a scientific designation, but also a symbol meant to inspire our environmental imagination as it conceptualizes nature.[4] We come to our concept of "nature" through the various ways we organize our experiences of the world that over time constitute what we mean by "nature." This formative process—our environmental imagining—is neither fundamentally given (though we certainly inherit typical organizing frameworks) nor value neutral.

In what follows, I suggest the Anthropocene symbolizes one way of engaging our environmental imagination—integrating our experiences of the world into a holistic pattern that gets performed as the intra-action of "human being" and "nature." As Forrest Clingerman rightly observes, "In deploying the idea of the Anthropocene, we do not solely understand the world through the presumed objectivity of scientific assessment, but equally as a reflexive emergence of *meaning* and *value* created by the intersubjective engagement between humanity and the material world."[5] The Anthropocene as a symbol—not merely a stratigraphic category[6]—reframes and guides the ways we conceptualize nature: What we mean by "nature" comes to include the human and the non-human. It proffers an ontological shift in understanding the relationship between the very structure of being human and the world that prioritizes the constitutive mutuality of their givenness to one another. At heart, the Anthropocene points toward a fundamental intra-action.

If we can grant this is the case, we need to consider (just as we might with other symbols, like the *imago Dei* in the previous chapters) two things. First, how does the meaningful ordering of our existence, which the Anthropocene proffers, synchronize with networks or families of other symbols that in a particular context of engagement will help clarify how this symbol communicates a participation in ultimacy?[7] The constructive aspects of this discussion will be the primary focus of the next chapter. However, here I will begin to explore how the constitutive mutuality the Anthropocene opens for our environmental imagination between nature and human being "corroboratively" taps into the intra-action of astrobiology and the refraction of the *imago Dei*.

Second, this chapter will explore more precisely how the Anthropocene has a disorienting effect on a seemingly well-oriented, familiar category: "nature." To do this, we must appreciate the flexibility in the use of the term "nature" through recent history in order to understand the extent of the Anthropocene's disorienting. This structure is in keeping with the account of symbols previously offered: To interpret a symbol adequately, we have to take seriously the history of its interpretation within some originating community. Here, that will be shaped in terms of a consideration of "nature" as it has been developed within a typology of American environmental imagining. Simply put, this chapter will seek to characterize how the story of human being and nature *inter*acting has slipped away in light of the call to recognize *intra*-action.

### Resonating with Astrobiology and the *imago Dei*

Since it taps into this "constitutive mutuality" between human being and nature, the Anthropocene fits well with astrobiology as its context of engagement and intersects with elements of the *imago Dei* stressed in the previous chapter. These connections represent a potentially deep and productive resonance. Before delving into the Anthropocene in its own right, it is helpful to try and gain some clarity about what harmonies emerge between astrobiology, the *imago Dei*, and the Anthropocene.

The Anthropocene expresses a *deep intra-action between human being and nature* that is in congruence with the intra-action between living-systems and the habitable environment crucial to astrobiology. To understand human beings as "geological agents,"[8] affecting the very physical processes of the earth, is to conceive of human being as a specific sort of living-system in intra-action with the habitable environment that is of interest to astrobiology. There is a coherence here between astrobiology and the Anthropocene: each pushes us to conceptualize human being in terms of how it *intra*-acts instead of how it *inter*acts with nature.

This connection may seem simplistic and broad, but it is important to note, nonetheless, because of the tendency to reduce the Anthropocene to a single issue. For instance, if taken as a "slogan" for climate change (as noted previously), then we can too easily imagine the Anthropocene as equivalent to climate change. This equivalency is an equivocation. Certainly, evoking attention to environmental crises—for which climate change should be of acute concern at present—is a crucial aspect of the symbolic power of the Anthropocene. However, the Anthropocene challenges the very externality of nature, Earth's systems, or our habitable environments in relation to human being.[9] Keeping intra-action at the forefront of our interpretation of what the Anthropocene means, as astrobiology necessitates, helps avoid misunderstanding the term as a simple catch-all for ecological disarray.

If the Anthropocene directs us toward constructing this shared history of nature and human societies, then it is in a sense immediately parallel to the account of the *imago Dei* as refractive offered in the previous chapter. Whether understanding the human being of the Anthropocene or the *imago Dei* of cosmogony, each symbol draws attention to our rootedness in wider environmental systems and our power to affect those systems. Further, both symbols imply that this power is so significant that it ought to be wielded with care and intentionality. There is a basic parallelism in how these symbols suggest we might meaningfully order our own existence in fostering the flourishing of these wider environmental systems.

More than being simply parallel, with astrobiology as the context of engagement, I am suggesting the *imago Dei* and the Anthropocene are *corroborative* symbols. I use the term "corroborative" to indicate that they help us explore the previously described tension between ultimacy and the concreteness of concern inherent to all symbols in mutually beneficial ways. Remember that as we primarily interpret a symbol in terms of its ultimacy, the symbol risks becoming meaningless through generalization; yet, if we primarily interpret a symbol as indexed to our existential concerns, it risks becoming meaningless through misappropriated concreteness.[10] Corroborative symbols mitigate this tendency, to become either meaninglessly abstract or unhelpfully concrete, in their pairing.

Broadly speaking, the refractive account of the *imago Dei* developed in the previous chapter is a symbol that tends toward ultimacy. It names the ground out of which we might construct an ultimate, meaningful existence: being a medium for the creative power of the divine that makes utterly new possibilities of meaningful existence available to creation. The Anthropocene, by contrast, is a symbol that tends toward the concreteness of concern. Interpreted as corroborative to the *imago Dei*, the Anthropocene can help pull this partner symbol back from the brink of abstracting vagueness. Vice versa, the *imago Dei* can help ensure that the Anthropocene does not lose its vitality and connection to imagining a meaningful order of existence by being mired in a taxonomy of heteronomous technical reason.[11]

In this tending toward concreteness, the Anthropocene is a symbol that *provokes* us into assessing our existential situation: It taps into how we conceptualize what is "of concern." In this case, the Anthropocene—as a symbol for the intra-action between human beings as geological agents who are never sufficiently understood apart from their affecting the habitable environment and being affected by it—continually points us back to how this mutuality must be made manifest in our daily lives. It turns us back to examine our mundane existence in light of the meaningful fulfillment promised in the encounter with ultimacy. Interpreted in corroboration with the Anthropocene, we can resist the tendency for the refractive *imago Dei* to become a platitude.

My intention is not to suggest that the *imago Dei* is connected unilaterally to the ultimate and the Anthropocene to the concreteness of concern. What I am stressing is that symbols exist in suites or networks by which they can corroboratively reinforce and constrain one another to better facilitate our participation in ultimacy. The *imago Dei* and the Anthropocene are good examples of symbols that can better facilitate such participation in ultimacy when interpreted together.

With this broad sense of how the Anthropocene can be interpreted in light of astrobiology and how it is a corroborative symbol to the *imago Dei*, we must consider through the rest of this chapter what is at stake in the concreteness of concern toward which the Anthropocene is continually provoking us. This prospect is no small matter because what we see all too clearly is a severe mismatch between the ideal of ultimacy and the concern of our existential reality. This mismatching stirs up a sense of anxiety indicative of an underlying disorientation. While we will examine disorientation in more detail in Chapter 8, it should suffice for now to point out that in disorientation things seem to slip; it is as though what we pursue recedes and evades our grasp as it is being pursued. It is "the moment in which what is within reach threatens to become out of reach."[12] Herein, an expectation is confounded and our world has to be rearranged if what continually slips away is to become available.[13]

Amitav Ghosh invokes this sense of disorientation, as it is relevant to the Anthropocene, in his work *The Great Derangement*. He
contends that in the face of the increasing extremity of environmental crises, we sense the limitedness of our individualistic accounts
of human agency. The autonomous individual of the modern age
appears insufficient in light of the crises facing us. In the wake of
this experience, we are confronted with the task of imagining new
ways to understand our being as it stands in relationship to these
non-human, environmental forces.[14] From here, the disorientations
only multiply: The disorientation of the Anthropocene calls out for
a politics that presses beyond appeals to the moralism of liberal individuals on which we have relied; it requires a renewed recognition
of non-human agencies banished by the mechanization of the modern age; and it reverses the expected order of modernity so that it is
not the technocratic elites to whom we look for a presaging of the
future, but to those on the margins as the ones who will experience
environmental catastrophes first.[15]

Jeremy Davies invokes a different sense of disorientation in his
work *The Birth of the Anthropocene*. He contends that in the face
of environmental crises such as species loss and climate change, we
have to find a way to meaningfully set human being into the history
of deep-time. As he observes:

> [T]he references to deep time bandied about in environmental
> news reporting are likely to be confusing and instantly forget-
> table for noninitiates. As one professor of geography wearily re-
> marks, "It is common when asking new undergraduates about
> periods of past time when things may have happened . . . to
> find a random selection of answers that fails to differentiate
> between hundreds, thousands, hundreds of thousands and even
> millions of years."[16]

The context of deep time is so bewilderingly huge—or disorienting—that references to it seem like a fantasy. The insights gained by
this hard work and scientific study take on the timbre of being even
less credible to my meaningful existence than something that happened long ago in a galaxy far, far away. In deep-time, my everyday
existence seems to inevitably become an insignificant blip. To dwell
with this disorientation means finding "some plausible way of coming to terms with the earth's bewildering antiquity."[17]

Ghosh and Davies provide two examples, but what various literature on the Anthropocene makes clear is that we feel anxiety in the face of environmental crises. (Even that we call them environmental "crises" is indicative of the anxiety attached to these events.) This anxiety is symptomatic of an underlying experience of disorientation: Environmental crises throw into question our way of meaningfully ordering our existence toward nature. The inherited, popular environmental imagination (for example, a passive nature ready for our use) and theological symbols (for example, the theodicies accompanying a robust, omnipotent theism) by which we could organize our meaningful existence in spite of the anxiety experienced in the face of these crises are insufficient. The Anthropocene, as a symbol, offers a different sort of hope and possibilities to deal with the anxiety through its persistent concern for intra-action, but we must work out how to shape our reality into a meaningful order of existence consistent with the hope of this symbol.

## Nature and the Anthropocene

It is in understanding how the Anthropocene throws askew traditional accounts of the relationship between human being and the world presupposed by an account of "nature" that Jedediah Purdy's account shines. He fills a critical gap in popular usage of the term "Anthropocene" by recognizing that "nature" has by no means been a static concept in the American imagination.[18] If the idea of nature is not static, but a concept like the Anthropocene is still a broadly disorienting and provocative symbol, we need to understand how across these various thematizations of nature there remains a common slippage.

Others have well traced the twists and turns of how this concept "nature" has been used across various and contradictory meanings.[19] We will not venture deeply into this tangled philological quagmire. Nonetheless, C. S. Lewis identifies a few helpful guideposts to bear in mind.

First, nature usually indicates some "class" or "kind." Nature designates that quality by which we might identify what it is like to be one thing and not another: It proffers a certain givenness or state by which something is. Such a state serves to contrast this thing of a given nature to those aspects of the world not possessed of such a nature.[20]

This sense of the word persists even when we use "nature" as a short-hand for the "natural world," but therein it takes on a second sensibility. As "natural world," nature is not only a given class, kind, aspect, or quality; it also implicitly indicates some sense of being undisturbed. The natural world denotes that class or kind containing those aspects of the world set apart from interference—with the assumption that what does the interfering is human being. With such a concept in hand, one further distinction can be made.

> We as agents, as interferers, inevitably stand over against all the other things; they are all raw material to be exploited or difficulties to be overcome. This is also a fruitful source of favourable and unfavourable overtones. When we deplore the human interferences, then the *nature* which they have altered is of course the unspoiled, the uncorrupted; when we approve them, it is the raw, the unimproved, the savage.[21]

Purdy implicitly recognizes that facets of these dynamics, which Lewis identified philologically, remain widely applicable to the term "nature" when he traces a genealogy of the American conceptions of this nebulous idea. The various ways we imagine nature, Purdy suggests, have always been connected to the ways in which we *value and inhabit* the natural world. In short, our imagination about nature has persistently, if at times tacitly, shaped a much wider and corresponding sense of ecology, economics, and politics in various periods of American history: What nature means has to be contextualized in terms of the wider worldview shaping our inhabitation of nature, thereby making it of value in particular ways. Nature is always taking on some set of "favorable and unfavorable overtones" as indicated in Lewis's wry observation mentioned previously.

Tracing this history of nature's valuation provides a crucial interpretive background to the ways in which we can make sense of the environmental crisis invoked by the symbol "Anthropocene." In fact, Purdy identifies the emergence of the term "Anthropocene" with a threefold crisis in ecology, economics, and politics, precipitated by the realization that a self-sustaining world must be built and preserved. Politics gets pride of place: building and preserving a self-sustaining world is fundamentally a political activity even if it clearly entails ecological and economic insights.[22] Moreover, this political realization corresponds to a dramatic shift in our conceptualization of nature with regard to the genealogy Purdy traces.

"Nature" has had many political meanings and alliances, as diverse as democracy and monarchy or hierarchy and equality, but it has always had one defining characteristic. In a purblind-ness that has marked all of human history before today, nature has been the thing without politics, the home of the principles that come before politics, whether those are the divine right of kings or the equality of all persons. *That purblindness is coming to an end with the Anthropocene.* The next politics of nature will be something different and more intense: an effort at active responsibility for the world we make and for the ways of life that world fosters or destroys.[23]

What Purdy's genealogy does is make clear why forming a poli-tics of nature is so difficult: Traditionally nature precedes politics—it is the natural canvas on which the artificial work of political painting can be pursued. Politics long denoted that realm or class of things that was wholly unnatural, that was cultural. It described the development of human social order and thereby entailed human interference—opposite to what Lewis identifies with the natural world. Politics is the realm of the interfered; nature is the realm of the undisturbed. A politics of nature is simply a contradiction in terms. Such an axiomatic contradiction between nature and politics as symptom for a hard divide between nature and culture has never really existed (and illustrating this is precisely the point of Purdy's genealogy).[24] Nature has always been political through its valua-tion, yet this has been subtended axiomatically as an impossibil-ity. If we take this inevitable politics of nature seriously—(dis)cov-ering it and noting that the world of which we are a part must be built and preserved—we need to make clear what is required by an "active responsibility for the world" while vigilantly ensuring this active responsibility does not inadvertently reinscribe the passivity of nature.[25]

In Purdy's work, the historical focus is distinctively American; however, the narrative he develops is one that has global implica-tions if the American context is taken as an analogue for the devel-opment of environmental reforms in other democratic contexts.[26] He suggests four versions of accounting for the American sense of nature: providential, romantic, utilitarian, and ecological.

The providential view emphasized both a sense of harmony with nature and a justification for the settlement of the American con-tinent. The proper settlement of nature reflects the harmonious actualization of nature's potential. The idea implied is an old one: The aesthetic relations of order in nature were evidence of a common pattern that could be raised to ever-greater perfection by human beings in morality and politics. The harmonious society was a perfect realization of the implicit harmonies that the American landscape manifested if one looked closely enough. The providential function of this wild terrain was to inspire the consummation of human development.[27]

Of Purdy's many examples for the providential view, perhaps none is as clear as his deployment of the jurist James Kent, whose legal arguments map out the political implications for connecting the potentialities manifest in nature's harmonies with societal development. Kent contended (despite the presence of Native American communities) that North America was legally empty upon European arrival. The key to this argument was that no one was making fruitful use of the land through cultivation; without cultivation, there was no ownership, only a usufruct claim to the land. Kent would go further, in line with the natural law theory of Emmerich de Vattel, to suggest that cultivation was operative as a legal tool because the act of cultivation represented the *obligation of nature upon human beings*. The providence of nature was cultivation by human beings that actualized the potentiality of nature's destiny.[28] Even the most beautiful and harmonious aspects of nature in its untouched observation paled in comparison to witnessing the cultivated development of a previously untamed wild space. The providential relationship of human beings to nature was the perpetuation and perfection of an implicit harmony of moral order.[29]

In some ways, the romantic inclination toward nature remains the most familiar. It is the obverse to the providential: a response to the emphasis on settlement and development that motivated the providential account. The romantic discovers in nature the beauty and harmony of which the providential is aware. Rather than insist, though, that this harmony points toward an unrealized potentiality of order (an order actualized through the domestication), the romantic view sacralizes the harmony of nature in all its beauty and sublimity. Politically, this sacralization entailed a need for preservation: Land cultivation and settlement disturb the harmony of nature.[30]

Here Purdy identifies a divide in the romantic heritage. On one side stands John Muir and the early work of the Sierra Club with the preservationist impulse. It focused on the development of public-land policies that opposed settlement in specific locales because the experience of unsullied nature provided an inherent source of self-knowledge. In the face of a soul-crushing society of endless activity, nature was a spiritual tonic that, in its immediate experience, imparted virtue to any who seized the opportunity to encounter it.

On the other side stand the transcendentalists—particularly Thoreau—with the impulse to encounter nature as an interlocutor. For the transcendentalist, the encounter with nature *could* be transformative. The transformation required cultivating a reflective attitude. Time in nature provided a space for questioning how people fit into the world such that we can contemplate how particular desires become of value to us, and we can subsequently engage in a moral revaluation.[31]

The romantic view, in both its preservationist and transcendentalist guises, understood that the harmonies of nature are not something human development perfects. They were instead a sacralized teacher: a balm to the anxiety produced by the cycles of production and activity that were part of cultivating nature into civil societies imbued by the providential view.

Employed in Progressive-era politics at the turn of the twentieth century, the conservationist approach is a cautionary recapitulation of the providential with attention to the harmonies emphasized by the Romantics. George Marsh's *Man and Nature* typifies this approach. Marsh provides a nearly apocalyptic account of what the consequences for shortsighted profiteering that threatens to diminish the natural health and productivity of our resources might be (with "resource" broadly construed to include not only "natural" resources but human vitality as well). An adequate way forward must meet both the social and moral dilemma ecological destruction posed.[32]

Educated management was Marsh's means to dealing with this dual problem. The aim was to generate sustainable interaction between human beings and nature through the *intentional* cultivation of natural resources. Importantly, though, this approach did not advocate some sort of repristination of wild spaces as though nature itself, apart from human being, evidenced its own most productive balance; in this sense, the conservationist approach is more akin to the transcendentalist than the preservationist impulse in the romantic approach to nature. A deliberate and managed disturbance of nature was the ideal: "'Conservation' meant imagining both natural and social life as complex systems that required educated management."[33]

Finally, the ecological approach forms an even more tense synthesis of the romantic and the conservationist approaches by emphasizing the interdependence between an organism and its environment. In this sense, "The bearers of ecological ideas combined the conservationists' description of nature with the Romantics' way of valuing it."[34] Specifically, the ecological approach values the robustness of interdependence.

From America's founding through the 1960s, the principle of managing nature, and political legislation related thereto, was one of "continental zoning." Tracts of land were given over to public management to effectively provide a means of planning and allocation. Zoning was a managerial approach directly in line with conservationist sensibilities. Yet the efficacy of continental zoning, as an approach that preserved the valued interdependence of the ecology, was being put under increasing pressure by human pollution.[35]

Pollution did not affect one zone to the exclusion of another; it threatened the very mediums that enabled and perpetuated the sorts of organism-environment interrelationship emphasized under the ecological approach. The spate of environmental laws from the 1970s reflect this awareness; they aimed to directly manage, through legislation, the mediums by which the interdependent relationships of the ecological approach emerge (such as air, water, toxic substances, and speciation) because no plot of land was immune from the interference of human pollution. Regulating these media, however, meant a profound shift in the managing of nature: "This meant environmental regulation of private industry and property, largely for the first time. An area of law that had mainly addressed the government's management of public lands now aimed at the environmental effects of the whole economy."[36]

Not only is this an important moment in shifting legislation beyond the management of public lands, but it also represents a critical departure from the romantic valuation of nature (whether in its immediacy as a tonic for the soul or in its more far-reaching potential to transform the reflective attitude of the individual). Now, nature has intrinsic value; it is not only *for* the assuaging of human urban anxiety or the development of a virtuous soul. Clearly, anthropocentrism remains in the ecological approach. There is a motive to preserve nature such that we, human beings, as one of many species might continue to live on this planet—that it continue to be habitable for us. Yet there is a clear shift; the pride of place given to the value pristine nature held according to its ability to communicate virtue to humankind gave way to a direct concern for the interrelationship of various organisms (besides humans) with their environment.[37]

With that critical shift in place, the way opened for an environmental economics that sneaks anthropocentrism back into our thinking just as quickly as it was removed. The driving impetus of environmental economics is to put a cost on the degradation of the resources and media that form the interconnecting lifeblood for the ecological approach to nature. The hope is that, through sufficient cost-benefit analysis, economists can price the act of pollution. This price would be an incontrovertible valuation of the full cost of producing goods that would cap the optimism affirmed by the boundless economic growth presupposed by the providentialist and conservationist approaches. Yet the shift to cost-benefit analysis makes these media at the heart of the ecological approach into yet another commodity at our disposal that can lead to questionable utilitarian calculations about current consequences and a possible disregard for long-term consequences in accounting for the full cost of producing a good.[38]

Purdy is clear that these four approaches do not provide a watertight typology; we adopt bits and pieces of each into our thinking about specific problems that we encounter wherein nature is a primary constituent. So what does tracing this history of the conceptualization of nature provide to enrich our understanding of the Anthropocene as a symbol?

It is important simply to note that we *can* trace a history of these conceptualizations of nature. We too often assume the Anthropocene coincides with an environmental awakening—as though previous generations simply did not understand what they were doing to the environment, but now we *know* better. Instead, we find a rich history of environmental reflexivity that we should not naively simplify. Each conceptualization of this reflexivity clearly recognizes there never has been nor ever will be a "nature" that precedes human interference.[39] Moreover, there is not one vision of nature nor a single sense of human being relating to it. Each approach is sensitive to what it perceives to be specific risks of environmental degradation characterizing its *Zeitgeist*.

We need to be more cognizant of this plurality of approaches weaved into social history as a tapestry for conceptualizing a relationship with nature predicated on mutuality, not mastery.[40] While studies in environmental humanities may be newly emerging within processes of college and university accreditation, they tap into a long pursued process of reflecting on the experience of relating human being and the wider environment. A sufficient account of the Anthropocene must not pose a radical break and should set itself into the context of this precipitating history. There is an important consequence to this line of thinking.

> The conclusion that forces itself on us, disturbing as it may be, is that our ancestors destroyed environments in full awareness of what they were doing. Industrialization and the radical transformation of environments that it caused by its string of pollutions went ahead despite environmental medicine; the ever more intensive use of natural resources continued despite the concept of economy of nature and the perception of limits. The historical problem, therefore, is not the emergence of an "environmental awareness" but rather the reverse: to understand the schizophrenic nature of modernity, which continued to view humans as the products of their environment at the same time as it let them damage and destroy it.[41]

In the Anthropocene, we must come to grips with the hypocrisy of a societal order that has recognized our deep connection to nature while nevertheless continuing to consume and destroy the natural world. Why have we so persistently ignored or marginalized the various means of chronicling our environmental reflexivity in our social and political decision-making? How can the symbol of the Anthropocene help stop such a process?[42] If the Anthropocene does not break from its precipitating history (as with the example of Purdy's genealogy), then will it not also continue the "schizophrenic nature of modernity" that understands the situatedness of human beings in nature but continues to rely on an account of human meaning making that backgrounds nature?

Criticism of the Anthropocene has largely come in two veins that express fear in the face of these sorts of questions: assuming that this new Anthropocene epoch only extends and intensifies the backgrounding of nature exemplified through Purdy's typology outlined previously.[43] First, critics contend the term "Anthropocene" reinforces the very binary separation of human culture—understood in terms of its technological development—from nature that it is intended to overcome. On one side of geological history stand natural processes. On the other side, we find this new epoch of humankind standing apart in its technophilia. In such a view, the Anthopocene represents an extreme version of the tendencies to cultivation advocated by the providential or conservationist approaches outlined previously.

Second, critics contend the broad-brush approach of the Anthropocene, in which we treat human being writ large as responsible for environmental degradation, pays insufficient attention to societal divisions. Surprisingly, if the first criticism likens the Anthropocene to the providential and conservationist approaches, this second criticism likens the Anthropocene to an extreme version of the Romantic and—to some degree—ecological approaches outlined previously in its concern to articulate a universalizing harmony that should exist between human beings and nature. The fear of this second group of critics is that when we universalize a pan-human culpability for our current environmental crises and homogenize the meanings and values with which we approach nature, we lose sight of the fact that not all human beings are equally geological agents in light of the worldwide markers being observed in the stratigraphic record. Human societies perpetuating a globalizing capitalism are geological agents. Yet we all too easily lose the nuance of this position in excoriating human culpability for our environmental crises.

Even if we give full credence to the complexity of the history of environmental reflexivity, wherein the simple narrative of nature as a background for human action cannot hold, Lewis's philological observation should remain ever-present in our minds. There is a powerful quotidian sense by which nature is understood to be the class of things free from human interference. The correlation that contrasts human meaning making with nature across Purdy's typology fits neatly with this quotidian sense of nature. Too often, we do not positively define nature; nature is merely a privation. Human interference in the quotidian sense of nature is then a cipher for making meaning and value. Nature is, by implication, that which is without meaning or value—the merely material background on which meaning or value may play itself out. A meaningful interpretation of the Anthropocene has to overcome this and vigilantly resist re-inscribing the gap between human meaning-making and backgrounding nature in its materiality. Insofar as we see a consistent tendency to inscribe the different senses of environmental reflexivity articulated at various points in history within a framework preserving human exceptionalism (in its contrasting the dematerialized vision of constructing meaning and value with the inert materiality of nature), any *meaningful* interpretation of the Anthropocene will be inherently disorienting.

Finally, this axiomatic shift is disorienting not only because it fundamentally reimagines the relationship between human being and nature, but also because such a reimagining throws into crisis the values inscribed upon nature that we have previously inherited. If "nature" is not merely a background to be inhabited, our current ways of valuing nature cannot exactly capture what is at stake in conceptualizing our relationship to the natural world. We have a crisis of values that arises when trying to reorient ourselves in a world where the straightforward division between human being and nature no longer holds: There is a vacuum of value vocabulary because those values previously available are premised on an untenable ontology.

This crisis and ensuing loss of actionable values is no trivial matter. Our values guide the ways in which we can fashion our lives into a meaningful sense of existence. To reach back to the previous chapter, having actionable values is critical to imagining how we might intentionally stand in the midst of creation as a refraction of the creative power of the divine. In such a situation, reclaiming symbols, especially doctrinal symbols such as the *imago Dei* with a long history of interpretation, is immensely important.

Symbols meaningfully orient us in the midst of the world by opening us to new ways of being-with one another. The long history of doctrinal symbols gives them a depth that allows us more easily to engage them and code the development of new values through them without the symbol simply dying from irrelevance. Doctrinal symbols have a staying power if we appreciate that their flexibility prevents any fixity or idolization in their interpretation (as was argued in the previous chapter). Taken this way, doctrinal symbols can provide a constancy that allows us to shift gently from one way of being in the world to another without the trauma of a radical break. By reclaiming symbols and dynamically reinterpreting them for changing contexts of engagement and new corroborative resonances, we create a buffer against the evacuating of our value vocabulary that comes with a shift in ontology. Doctrinal symbols, as presented in the previous chapter with the *imago Dei* as an example, can be a tool used persistently and critically to re-orient our values in deference to shifting ontological frames of reference.

What Purdy's reflection on the Anthropocene aptly adds to this contention is that "the range of interests one can pursue forcefully and effectively in politics has much to do with the specific ideals

different set of imperatives by which to structure our sense of human being in the world.[8] In terms of the wider argument in this book, planetarity is a sort of intra-action. It makes the entire network of human being and the wider habitable environment into a primordial category from which human being can never be wholly removed. Spivak slyly indicates as much, noting that by suggesting that planetarity is in the "species of alterity," she had in mind an alternative to the global being in the "species of eternity."

The imperative of planetarity involves fundamental alterity: What it means to be human can be manifest only through our being intended toward the other. Planetarity indicates something further, however: Alterity grounds us. As our ground, or perhaps better would be to say "our abyss," this alterity is not derived from us—it is the figure that shapes our meaningful experience of existence while wholly exceeding us. It is "the (im)possibility of underived intuition."[9] When we understand ourselves to be planetary creatures instead of global agents, this alterity is self-evident: warranted by being given over to us in all our experiences while itself never being something grasped by our subjectivity.[10] Spivak's formulation emphasizes that alterity is something given over to us so that our being intended toward the other is never interpreted as a willful act of some primordially self-sufficient subjectivity.

More practically, we might say that planetarity is a disposition toward infinite questioning. It begins by questioning if and how inherited global ways of knowing inflict a conceptual violence on the ways we can give an account of our experiences: squeezing what we know into stable, fixed categories by which such experiences can be rendered meaningful. Giving attention to those experiences and identities that are abnegated by the process of fixed meaning-production characteristic of the static regime of global thinking, planetary knowing is performed as an iconoclastic alternative. The persistent questioning characteristic of planetarity becomes a means to eschewing supposedly stable boundaries of any assured identities relied upon by global knowing—such as "human being" or "nature."

Instead of proposing some series of alternative, but nonetheless fixed, visions for identity construction, in planetarity we engage in an aesthetic reimagining of the boundaries between the self of human beings and the other features of the world. This is a means of co-creating alternative futures and a plurality of planetary identities, where previously there was a single identity in global knowing, which recognizes the *embeddedness* of human being. No longer can we rest the stable identity of human being on top of an instrumentalized nature that supports its flourishing; the identities that come to flourish on a planetary way of knowing are holistic and blur the boundary between human and non-human others.[11] Here an opening re-emerges on what the connection between planetarity and our environmental imagination might be.

In planetarity, recognizing our grounding in alterity that is irreducible to our subjectivity, we are given over to ourselves in a new way. It is a way that is concurrent to the Anthropocene vision of human mutuality that comes with being a *participant* in the deep history of the planet instead of the master over the value attributed to global exchanges.[12] This participation is crucial.

To be a planetary creature is not only a rejection of a scientistic ideal by which human being acts through taking a view from nowhere—outside and above—on the world.[13] It is this, but it is also more. Planetarity is also an affirmation of *being* provisional and positional. In the planetarity of the Anthropocene, we *are* through our dependence, not despite it. Our being cannot dissolve these dependencies; technically speaking, we have our existence only through a recognition of the in-between space that is maintained in the non-separable relations of alterity.[14] But, to recognize the primordial quality of this in-between space of alterity requires a persistent suspension of self-interested appropriation of global exchange in order to engage the mutuality of planetarity—to be responsible to others in the alterity of my planetary being and to be responsible for engaging in the mutuality of participating in the deep history of the planet.

Adopting such a shift, for which Spivak and others advocate, makes a tremendous difference. If in planetarity I am made responsible in the alterity of being intentioned toward the other *and* this alterity utterly exceeds me as my grounding condition (i.e., it is not reducible to my subjectivity but a condition of my having meaningful experience in the first place), then to act responsibly is a fundamental aspect of the human condition. Responsibility is not an obligation or a duty laid upon me as a result of being human. It is more fundamental than that. Responsibility becomes a human right: Its practice is an index for a human way of being.[15] Without responsibility, we become something less than human; but this is not some hyper-individualism. My responsibility is never mine alone. Responsibility arises in alterity: It is given over as *ours*, expressing the mutuality and dependency of always being in-between.

To effect such planetary re-figuring, Bauman argues we must begin by destabilizing foundationalist accounts of nature that can all too easily make recourse to global knowing.[16] This destabilization is a "critique," in the philosophical sense that it must contest the assumed contradiction by which nature's identity is established and imagine instead how the performance of a planetary process shifts the habits and systems of nature such that human being intra-actively participates in this ongoing process. Nature, in its planetarity, must be "a multiperspectival emergent process" that includes "humans, cultures, religions, ideas, imagination, atoms, ecosystems, the earth, the universe, and all other levels of reality."[17]

More than once, Bauman describes this as a *natura naturans* without the *natura naturata*.[18] He is emphasizing that the process of meaning-making is so radically immanent to the variety of ways that "nature" comes to be that there is no fixed or absolute position from which to identify "nature" in contrast or opposition to some transcendent point apart from the process of nature naturing. Meaning-making is immanent to nature, not something inscribed upon it or rendered outside it. There is no point outside nature from which this inscription could occur.[19] Instead, there is perpetual "nature naturing" without a fixed point toward which this naturing is directed; nature is not some-thing against which we are set, but it describes the multiple flows of planetary becoming.

It should come as no surprise that this refiguring of nature entails a correlate refiguring of human identity in light of a planetary framework that often invokes the philosophical concept of "enfolding horizons."[20] However, conceptualizing this enfolding is very difficult. Bluntly put, the enfolding of horizons describes a relation of non-separable difference enacted through performativity or repetition. The idea of enfolding and unfolding requires doing away with any conventional notion of the self as a static center from which this process of folding begins. Instead, the fold is a moment or an event from which identity itself tentatively emerges.[21] Our identity unfolds this relationship of non-separable difference to our horizon as a more primordial unit of thinking than "self" or "horizon" understood apart from each other.[22] We then again and again enfold horizons anew, though never precisely in the same way. The enfolding is never precisely recapitulated as we light upon new horizons.

We are not isolated individuals who enter into relational or dialogic ways of knowing or being in the world: We are fundamentally relational—not a self that enfolds horizons but the enfolding of horizons emerging as a self. The self and the horizon form a relation of non-separable difference. To be the embodiment of this folding is to be a de-centered self: The folds do not happen to an otherwise static entity; the folds comprise our very sense of self—a self only made static through the repetition of enfolding and unfolding particular horizons.

Moreover, environmental thinking ensures that this intertwining of alterity at work in the non-separable difference of the enfolding and unfolding of horizons is not merely an interhuman phenomenon. Human being is a complex layering of enfolding and unfolding occurring across multiple levels of scale simultaneously.[23] We are microbial communities just as well as participants in wider ecosystems.[24] The self-world distinction cannot be propped up unproblematically as though we can draw a sharp boundary at the skin, establishing a dermal metaphysics that unproblematically separates us off from the world around us. Our bodies are themselves, to the very core, complex enfoldings and unfoldings of the wider world in which the many beings on this planet—the whole plethora of nature naturing—form a part.[25] As Bauman argues, "At the very center of the self are multiple earth others."[26]

It is only from this conception of a de-centered self, a multiple subject, an enfolding and unfolding of horizons, or a post-dermal metaphysics (whatever term we might prefer) that a truly planetary sense of human being and nature can develop. In what follows, this is the tradition that I am invoking continuously and implicitly. Such an identity is one that emphasizes this grounding, wherein we intentionally extend such an awareness of multiplicity into the performances and repetitions that give shape to our sense of self and world. In short, if we are truly a multiplicity in our subjectivity then it should follow that our epistemologies and conceptions of agency should be similarly multiple.[27] To affect such a shift, fully, planetarity requires a re-imagining of ontological categories that can account for the layering and multiplicity of the de-centered self.

As a first step in this direction, Bauman suggests the epistemic and agential multiplicities that need to be accounted for by a planetary identity or planetary knowing rely upon us "taking on" (as in both trying out and consciously considering) the abject into our self-understanding. He makes the case for the performance of planetary identities that respond (both epistemically and ethically) to the abject identities we seek to abandon through performance.[28]

These multiperspectival planetary identities are then responsible to and responsible for a flourishing of the world that goes far beyond a concern for mere human flourishing. What is needed is an ethics of planetary flourishing—a polyamory of place—that empowers more complex accounts of environmental action and challenges any remnants of place-based approaches to nature (i.e., nature is the setting of standing reserve for exclusively *human* flourishing). These place-based approaches deny the chiasmic, ecological intertwining that is part of the porous border between subjects and many planetary others, whether human or not.[29]

To do this, though, requires a thorough consideration of how a bodily mood affects our interpretation of nature.[30] Even as our sense of self is de-centered, we have an experience of nature through the integrity and intentionality of our bodily experiences that are always given over in a mood that characterizes our attunement to the world around us. Bodily mood renders the possibility of any particular experience of nature. If in the Anthropocene we are planetary creatures, living out such an aspiration requires attention to fostering moods that will allow us to discover how *planetary* flourishing is "of concern"—not merely human flourishing or a falsely conceived sense of individual flourishing. Further, we must locate human being (and honor this location) within the complex multiplicity of planetary subjects that can foster new ways of becoming a planetary community. As *one of many actors with agency* in planetary communities, our performative habits must reflect this breadth and be driven by care for the flourishing of these other modes of agency—a care that ought to lead us to question our reliance on understandings of modern liberal individualism in our political, economic, and legal systems.[31]

Given that the Anthropocene contends we are planetary creatures, we might summarize that this is disorienting in at least three ways. First, we are not our own. In planetarity, we are under the "species of alterity"—a grounding alterity that is irreducible to my own subjectivity. Old conceptions by which nature is the limitless background for realizing meaningful human ideas cannot hold in the Anthropocene.

Second, this grounding in alterity places a host of non-separable differences at the center of who I might understand myself to be. *I am utterly dependent.* Yet with this dependency comes responsibility—responsibility not understood as an onus or obligation fulfilled beyond the satisfaction of our own needs, but responsibility as a fundamental right characterizing a human way of being in the world. This is certainly not meant to imply that responsibility or dependencies are equally distributed in the Anthropocene. Clearly, our habits of consumption and energy use indicate this differentiation. But if these are fundamental characteristics of human being, they must be given a primacy of place in thinking about our place in the Anthropocene.

Third, we need an ontological vocabulary that better accounts for the non-separable relations of intra-action at work across levels of scale and complexity to realize the extent of our dependency and the breadth of this grounding in alterity. Planetarity challenges our ways of thinking about ourselves as though our bodies and the borders of our skin are static and impermeable. Thus, as planetary creatures we must take stock of the moods by which we are attuned to nature. We must consider more intentionally what moods ought to be fostered in order to affect experiences of nature that give credence to our intra-action and might resist the tendency previously exhibited to ignore accounts of environmental reflexivity.

Yet this is only half of the way that the Anthropocene is disorienting. Not only are we planetary creatures in the Anthropocene, spatially distorting the relationship between human being and nature, but we are also thrust into deep time. It is to this temporal sense of disorientation that we must turn now.

## Beyond the Anthropocene Epoch: A Sapiezoic Eon?

As Jeremy Davies well recognizes, the current environmental crises we face have thrust human beings into deep time. He observes,

> [P]oliticizing deep time is a habit peculiar to environmentalism. Ecological politics struggles with the difficulty of imagining the distant past. Efforts to fix geological time in familiar, tangible terms only make it even stranger. . . . Certainly, the geological timescale can be learned: Geology students need to do so in order to pass their exams. But it is hardly reasonable to make that memorization exercise a requirement for ecological good citizenship. Everybody is already overburdened with the weight of information available to them about the state of the planet. What is needed instead is some plausible way of coming to terms with the earth's bewildering antiquity, now that climate change and species loss have forced the subject forward to public attention.[32]

Like it or not, as geological agents who, through our human societies, have come to leave our stratigraphic mark in the layers of the Earth by affecting the climate and wider environment, we are inevitably a part of the Earth's story. This is a story bewilderingly longer than that of human history. The distant past is set at a time immemorial from the rapid pace of human history.

As the historian Fernand Braudel noted, there are three temporalities in history: "that of nature and climate, almost immobile and not affected by human action; the slow temporality of economic and social facts; and finally, the rapid temporality of events vibrating to the rhythm of battles, diplomacy and political life."[33] There are exceptions to this way of telling history, particularly as with early environmental historians such as Aldo Leopold. By giving history from the perspective of non-human forces, he brought to the foreground the seemingly "immobile" temporality of nature as a consciousness raising activity.[34]

In turning toward the temporal disorientation of deep time, astrobiology might explicitly provide a helpful frame of reference for figuring the Anthropocene (or be at least complementary to stratigraphic approaches). It does this by changing the scale at which the problem of environmental crises symbolized by the Anthropocene are considered and in so doing affects an even more drastic de-centering of human identity than that imagined in terms of planetarity's grounding in alterity. Astrobiology takes alterity to an entirely different scale. The Anthropocene seen from astrobiological eyes provides a vista from which to view the emergence of human being from wider biohistorical systems. At such a vast scale, the typical framing of the disorientation of the Anthropocene—in terms of the relationship between "human being" and "nature"—is itself dwarfed, and this dwarfing is itself disorienting.

Perhaps no one has made a more accessible and sustained case for thinking about the intersection of astrobiology and the challenges of the Anthropocene than David Grinspoon.[35] He develops a novel position concerning, what he considers, the fundamental Anthropocene dilemma: "[W]e have achieved global impact but have no mechanisms for global self-control."[36] He takes it as his task to consider how humanity might move forward to manage the future possibilities of the planet wisely, working from an astrobiological perspective that intentionally thinks of the Anthropocene in terms of a planetary time scale and considers how to frame our thinking through the intra-action of habitability and living-systems.

On the one hand, this planetary management sounds exactly like the sort of technophilic arrangement that some authors are concerned will cause us to ignore the possibilities of planetarity outlined previously, particularly regarding alterity, as a conceptual grammar of environmental reflexivity for the Anthropocene.[37] However, the emphasis on astrobiology and its intra-action in Grinspoon's proposal should serve as a constant check against marginalizing our grounding in alterity. Whatever sorts of "global self-control" might be developed, they must reflect a thorough understanding of the ways in which a habitat for life emerges on the Earth. By taking this perspective, Grinspoon is integrating the Anthropocene into the geological history of the planet in its entirety. The significance of deep time in Grinspoon's proposal, then, is quite different from Davies's use of the concept even if they share a fundamental recognition of the disorienting effect of deep time; the difference, I would suggest, has to do with the astrobiological context of engagement.

Davies is concerned with making deep time comprehensible to the wider public so that we understand the contemporary ramifications of climate change. To do this, he focuses on setting this epoch within the context of the Phanerozoic eon and into the specific shifts constituting the Quaternary period of the Cenozoic era: from the late Pleistocene epoch, through the Holocene epoch, and into the Anthropocene epoch. He is not concerned with the far future of the Anthropocene, which he fears detaches us from the present political implications of the environmental crises we face.[38]

By contrast, Grinspoon argues that we need to understand the very formation of the climate on planet Earth if we are to appreciate the distinctiveness of the Anthropocene epoch—requiring us to have a much wider perspective on deep time extending into both the past and the future. Thus, by turning to astrobiology, Grinspoon is affecting a style of environmental consciousness-raising that is akin to that of Leopold. Instead of making mountains or prairies the subject of historical narrative, though, the planet itself in the wider nature of the cosmos becomes Grinspoon's subject.

Perhaps rather than Leopold specifically, Grinspoon's project most clearly resonates with a short documentary film from 1977 called *The Powers of Ten* in turn adapted from the idea put forward in Kees Boeke's 1957 book *Cosmic View*.[39] The idea is that the audience's view of a typical day at a park in Chicago zooms out (from one square meter) and then back in. The viewer quickly glimpses the importance of thinking at various levels of relative scale. If

Davies's work continues to think about the Anthropocene and climate change at the immediate scale of human experience (making deep time accessible to human experience), Grinspoon is zooming out to think about the issues of deep time more directly at a planetary scale and only subsequently considering how human being fits into this widened frame of reference.

> The planetary perspective allows us to step away from the noise of the immediate present, to see ourselves from a distance, in time-lapse. When we do so, what we see is not just a problem facing our civilization but an entirely new evolutionary stage in the development of life. In seeing ourselves as a geological process, we also see the planet entering a phase where cognitive processes are becoming a major agent of global change. Earth's biosphere gave birth to these thought processes, which are now in turn feeding back and reshaping its changing planetary cycles. A planet with brains? Fancy that. Not only brains, but limbs with which to manipulate and build tools. We are just beginning to come to grips with this strange new development. Like an infant staring at its hands, we are becoming aware of our powers but have not yet gained control over them.[40]

For Grinspoon, the Anthropocene is of significance in the history of the Earth. That current human behaviors are leaving an indelible mark in the geological history of the planet remains critical to his account, but he wants to draw our attention to something broader: what the *legacy* of the Anthropocene might be in the Earth's ongoing history. More specifically, this is to ask if the Anthropocene represents a planetary event, a planetary epoch, or a planetary transition to a new eon. The difference between these terms is significant. A planetary event marks a disruptive and short-lived impact in planetary history. By contrast, an epoch indicates a prolonged influence of change on planetary history. A planetary transition to a new eon, however, marks something extreme; this would mark a transformation on the scale of the emergence of life or the evolution of complex multicellular creatures.[41]

While most people believe that it is too grandiose to think of the Anthropocene as introducing a new epoch, Grinspoon is suggesting that from a cosmic perspective it might be too modest to think of the Anthropocene as merely an epoch: "A shift to a new epoch is not that rare. An epoch typically lasts for a few million years, which,

for Earth, means it's no big deal. Yet this is not merely another geological shift among many in Earth's long, ever-changing history. The
advent of conscious agency as a force of change on Earth is a major
inflection point in the way the planet functions."[42] In this spirit,
he suggests thinking of the mature Anthropocene (a term we will
return to shortly) as ushering in a fifth eon in Earth's history.

Remember that from an astrobiological standpoint, the eons of
the Earth mark fundamental transitions: these so-called "inflection
points" where the systemic operations of the planet undergo a major shift. Grinspoon, admittedly all too simply but with astonishing
conceptual clarity, lays out a correlation of the four eons and their
respective transitions. The Hadean eon marks the very origins of
the Earth. The Archaen eon marks the origins of life. The Proterozoic eon marks the advent of a biosphere. Finally, the Phanerozoic
eon marks the origin of complex, multicellular life.[43] With the passing of each eon, a new, intra-active relationship between life and
the planet, between living-systems and the habitable environment,
develops on a cosmic evolutionary scale.

If the Anthropocene is ushering in a fifth eon in Earth's history—
this new intra-active relationship between living-systems and the
habitable environment—Grinspoon proposes calling this the Sapiezoic eon. The change he envisions during this eon is on par with
those from the previous eons: a point when "cognitive processes
become a stable part of a planet's functioning."[44] If we wanted to
draw a sharper distinction from the preceding Phanerozoic eon,
we might say that the Sapiezoic eon would be one in which a catastrophic event (a technical term we will examine in a moment)
did not result in massive extinction—a cycle that punctuates the
Phanerozoic eon.

Importantly, Grinspoon is careful to emphasize that *it may not
be humans* that live into the wisdom of planetary management suggested by his Sapiezoic eon. We are not the pinnacle of creation;
there remains a careful Darwinian hedging that continues to understand human beings as evolving creatures participating in the unfolding history of the planet that distances Grinspoon's work from a
more teleological refashioning of cosmic evolution as with Teilhard
de Chardin. Human being is not triggering the occurrence of the
noosphere as an omega point;[45] a Sapiezoic eon is not anticipated to
be the last or culminating phase in the history of the earth, *nor is it
certain that homo sapiens are the species that might induce such
an eon.* There is no guarantee, in terms of the cosmic evolutionary

scale of the planet, that this experience of the Anthropocene will represent anything more than a mere planetary event—perhaps at best inducing an epoch correlated to the sixth mass extinction of the Pleistocene epoch instead of the transition to a new eon.[46] The Anthropocene might then be a major planetary event within the wider context of the Phanerozoic eon: a part of the planet's functioning, but hardly a stable part.[47]

While Grinspoon may be shifting our attention to the Earth's history, its history remains connected with human history. Can we be the "intelligent" life ushering in a Sapiezoic eon? How do our actions wrapped up in the "rapid temporality" of historical and political events make our continued survival and the opening of a new phase of planetary history more or less possible?

To get at these issues, Grinspoon suggests we must understand why the climate of Earth has remained relatively steady in comparison to other planets. Without this steady climate, the continuous planetary habitability of Earth certainly would be in question and ideas about the Anthropocene (whether understood as a planetary event, epoch, or transition to a new eon) would be mute. Understanding how this steady climate arises and what its relationship is to the study of planetary habitability is where astrobiologists can engage contemporary questions concerning the Anthropocene, particularly as it is symbolically associated with climate change. The questions of how living-systems and habitable conditions begin intra-acting and might further flourish (forming the conditions for continuous planetary habitability described in Chapter 1) is at the heart of many segments of astrobiological research—and the study of the Anthropocene is deeply concerned with how life might continue to flourish (or how it will fail to flourish) in light of climate changes and shifts in the habitable conditions of the planet so drastic they are preserved in stratigraphic layers.

Using comparative planetology, and focusing particularly on Venus and Mars, Grinspoon suggests a taxonomy of planetary catastrophes by which to guide our thinking about the Anthropocene.[48] Catastrophe is used in the technical sense of a sudden and dramatic change in the state of a planet—it is not meant to indicate any judgment about the type of change being made.[49] His taxonomy includes four types of catastrophic change: random, biological, inadvertent, and intentional. Even in the names, there is a significant divide between the first two types of change and the final two types as regards issues of agency and intention. In the case of random and biological catastrophes, issues of agency and intention are absent; catastrophic changes of this sort simply happen "unwittingly." By contrast, inadvertent and intentional catastrophic change do intersect with issues of agency, intention, and to some extent awareness.[50]

Now, why is it that Earth's climate has remained relatively steady while the climates of Venus and Mars have demonstrated run-away warming or cooling? Even in light of comparative planetology, the formation of Earth's climate is a bit of a puzzle. Given the faint young sun, that our sun would have only been about 70 percent as bright as it is now at the time of the formation of the Earth, we would expect that the Earth would be completely icy. There is no definitive account of why this is not the case and it is still a topic of debate.[51] One line of reasoning suggests early in Earth's history a negative feedback system (purely geochemical—a product of random catastrophic changes in Grinspoon's typology) developed as a sort of precursor to our current understanding of the carbon and water cycles (*bio*geochemical—a product of biological catastrophic changes).[52]

In short, as volcanoes added carbon dioxide to the atmosphere, weathering reactions sequestered atmospheric carbon dioxide in carbonate rocks; these rocks were then subducted and the captured carbon was re-released through volcanic activity. This is a *slow* process, but what is curious is that weathering reactions are sensitive to climate. In warmer conditions with more rainfall, carbon dioxide is more quickly sequestered from the atmosphere. This reduction in greenhouse gas leads to cooling and eventual freezing in which the carbonate rocks are buried beneath solid ice. Under the ice, the weathering reactions severely slow down, or cease altogether, and so carbon dioxide begins to build up in the atmosphere again. Even as the sun grew brighter and warmer, this interaction of volcanic and hydrological cycles self-adjusted and thereby moderated greenhouse gases.[53]

One of the advantages of such a model is that it can also account for why our planetary neighbors did not develop like the Earth (despite at times being in the habitable zone of our sun). In each case, some aspect of the interaction between the hydrologic and volcanic cycles broke down. In the case of Venus, the hydrologic cycle likely broke down; water vapor is a greenhouse gas and, as it warms the atmosphere without condensing into rain, it causes more heat to build up, which leads to greater water evaporation and, in turn, more atmospheric heat.[54] By contrast, in the case of Mars it was likely too small to sustain active plate tectonics driving the volcanic activity to replenish atmospheric carbon dioxide and a magnetosphere to protect it from solar winds.[55]

What comparison to Venus and Mars forces us to consider more deeply is at what point does the interaction of geochemical cycles become significantly *bio*geochemical? When do we cross a certain threshold by which the influence of life on a planet becomes crucial to maintaining the geological and atmospheric conditions evidenced by a planet? Venus, Earth, and Mars, may have started out quite similarly, but *life on Earth*—as opposed to Venus and Mars—became a major contributor to the long-term physical dynamics of the planet itself.[56]

On Earth, biological processes become integrated into these geological processes—they become biogeological processes. Life, then, works a bit like a generator kicking on if the power goes out in your house; biological processes can keep things steady for a while until the long-term system, the geochemical cycle, is square again. It is not as though the biological processes kick in only when something goes haywire with the geological ones. Nonetheless, the redundancies that life builds into the system are key. If the geochemical cycles broke down in some way (say for instance not sequestering sufficient carbon dioxide eventually trapping enough heat so that water vapor no longer condenses and only exacerbates the climactic warming), then the biosphere is a redundancy that can help steady things for a while and in the long run facilitate the persistence of far from equilibrium conditions on the planet. In a sense, the biosphere steadies the planet against random catastrophic change.

There is an intimate correlation, an intra-action, between life and climate—or we might say a non-separable difference forming habitable conditions to borrow the language of planetarity. Life is not merely an afterthought to the formation of fortuitous conditions. Once minimally stable and sufficiently populated (the emphasis here is on continuous planetary habitability after all), we can recognize the pervasive influence of life. The climate formed on Earth is not a pure contingency upon which life occurred as an afterthought to fortuitous conditions: "In short, life has *always* been a geophysical force; equally, the geology of the Earth, unlike that of Venus, has been influenced by the laws of biological evolution for an inordinate length of time."[57] The Anthropocene is not distinctive because suddenly living things are geological agents; that has long been the case.

Recognizing this, Grinspoon advocates for a shift: "Perhaps life is something that happens not *on* a planet but *to* a planet: it is something that a planet becomes."[58] There is a difference between living worlds and non-living worlds; a planet becomes biologically distorted or modulated to such an extent that "living" as a descriptive moniker does not apply just to an organism or a system of organisms but to the unity of living-system and habitable environment as a whole. Life describes a particular physical state of the planet that promotes the maintenance of a far-from-equilibrium system:[59] a stabilization of the random catastrophic change producing geochemical cycles that facilitates the formation of a habitable environment. With such stability, life can evolve—exploring possibilities and ways of being in the midst of this stabilized, habitable environment that are not possible if it is continuously subject to random catastrophic changes not being kept in check by the various negative feedback cycles.

Yet life is not merely a mollifying force for the habitable environment; it can invoke its own sort of catastrophic change—the second of Grinspoon's four types. Consider the great oxygenation event of around 2.4–2.1 billion years ago. For hundreds of millions of years, cyanobacteria (blue-green algae) proliferated in the oceans and harvested energy from the Sun through photosynthesis. The byproduct was, of course, oxygen. Initially the oxygen was soaked up by iron in the Earth's crust and other atmospheric reductants. In this sense, the produced oxygen could be absorbed by mechanisms ready-to-hand within the existing biogeochemical cycles of the planet. Eventually, however, the crust could not be oxidized any more. When this combined with a decrease in atmospheric oxygen consumption from other possible reductants, the amount of atmospheric oxygen rose dramatically and quickly; nothing checked the photosynthetic production of oxygen by the cyanobacteria, but there was not any biogeochemical cycle to sequester the oxygen and prevent its continued accumulation. The buildup of oxygen would have been highly corrosive and fatal to most species of life—oxygen violently reacts with most organic molecules when not tempered by the process of respiration that stores such energy in phosphate bonds.[60]

Cyanobacteria nearly wiped out life on Earth through the runaway production of oxygen. As we saw on Mars and Venus in terms of random change, here we have an example of dangerous, disruptive, runaway positive feedback as a biological catastrophe. Eventually, the atmospheric shift described could have been very bad for the cyanobacteria themselves. A great freeze happened on the Earth in the wake of this oxygenation event, likely due to a significant reduction in methane gas providing a greenhouse effect, which could have simply stopped the occurrence of all life on Earth. Yet life persisted as some existing feedback loops (volcanic emission of carbon dioxide and other greenhouse gasses) and newly evolved loops to prevent runaway positive feedback (cellular respiration) contributed to a climate rebalancing.

There is an important feature of catastrophic change that the great oxygenation event makes clear: Once a catastrophic change takes hold, it is not easily rolled back. Once the atmosphere was oxygenated, there was very little that could be done to lessen this state. Instead, new forms of life evolved to contribute to the climate rebalancing; specifically, it was large, complex, multicellular organisms that evolved (think of all the familiar types of plants and animals we might usually imagine in our everyday lives when thinking about living things). Despite subsequent mass extinction events from random catastrophic changes, complex multicellular organisms have persisted. Catastrophic changes leave irreversible marks in the geological history of the planet.

This is important for our thinking about the Anthropocene as well. If we have crossed into the Anthropocene, then part of what this needs to symbolize is the death of the Holocene.[61] We cannot nostalgically look back to days before climate change any more than we could look back to the days before the great oxygenation event. Those changes have occurred and are occurring. What we can do is look forward to what the impact of the Anthropocene might be: Will it be an event, an epoch, or an eon? It could be any of these.[62] Yet to imagine which it might be, we need to think about the implications of a last form of catastrophic change.

At this point, I would suggest a slight amendment to Grinspoon's typology. He proposes two final types of catastrophic change: inadvertent and intentional. Inadvertent changes are "planetary-scale changes that result from intentional applications of technology but are not themselves intended."[63] They combine a technological prowess yielding global impacts with disconnection from our personal agency and intentionality given the scale at which the effects of these technologies are manifest. Intentional catastrophic change implies "Global-scale changes that result from actions undertaken with forethought by an intelligent technological species aware of its actions and their consequences."[64]

Forethought is crucial to distinguishing inadvertent and intentional catastrophic changes in Grinspoon's typology: Was the change an accident or undertaken as an intentional form of management. Instead, I would suggest that we emphasize a common feature of Grinspoon's accounts of inadvertent and intentional catastrophic changes and put forethought aside for the moment. If a catastrophic change represents a sudden shift in the state of the Earth with consequences that cannot be easily rolled back, then it is the advent of *technology* that is crucial. Beyond random and biological catastrophic change lies technological catastrophic change.[65]

Much like with biological catastrophic change, it is not just a single organism or group of species that is significant here. Biological catastrophic change occurred when a living-system, a biosphere was integrated to the geochemical cycles of the planet. Here, too, it is not a single or even suite of technological developments that is indicative of catastrophic change. Rather, it is the rise of a technosphere (to be parallel with the biosphere) and understanding how it is integrated to the biogeochemical cycles of the planet—such that they become *techno*biogeochemical cycles—that marks this catastrophic change.[66]

Davies provides a helpful example with regard to his emendation of the carbon cycle:

[C]arbon must pass through multiple phases in the course of its cycle. Traditionally, they can be grouped into biological and geological phases, but the carbon cycle now also includes a novelty: an *economic* phase. While fossil fuels are still underground, they become assets of the companies that own them. This economic phase is an essential part of the modern carbon cycle. If the reserves did not have economic value, labor power could not be mobilized to bring them to the surface; the quantity of value that they possess determines whether and when labor power is attracted to them.[67]

If we understand economics as a specific sort of technology in the technosphere (a case I will not make in-depth here, but seems reasonable in light of various approaches to history in the Anthropocene),[68] then fossil fuel extraction provides a ready example of catastrophic change insofar as the technosphere is being integrated into an existing biogeochemical cycle. Imagining where and how technologies are being integrated to other existing biogeochemical cycles is crucial to understanding the Anthropocene if it is being marked off by this sort of technological, catastrophic change.

To repeat, the concern is quite specific. In both random and biological catastrophic change, sudden shifts occur in conjunction with some sort of runaway, positive feedback as a result of the global level event. In each case, the positive feedback is eventually mitigated by its inclusion with a negative feedback cycle. The question, then, is really one of how a technosphere sets off some form of positive feedback by not being well or sufficiently integrated to existing biogeochemical cycles.

Now we can return to the issue of foresight that Grinspoon indicates is the differentiating point between inadvertent and intentional catastrophic change in his typology. If both of these are forms of technological change, identifying what difference foresight might make ties into his hope that the Anthropocene is more than a geological event or an epoch. If technological change proceeds inadvertently, then the Anthropocene seems likely to be an epoch that, given the wider history of the Phanerozoic eon, will result in another mass extinction that—for all intents and purposes—may not so much represent a new form of catastrophic change, but a variation on biological catastrophic change.[69] This is how I would characterize what Grinspoon refers to as the "proto-Anthropocene."[70]

However, if technological catastrophic change is induced with foresight toward the consequences of these changes, then we enter into what Grinspoon terms the "mature Anthropocene" that correlates to his hoped for Sapiezoic eon.[71] Herein, humankind moves past having erratic planetary effects and can become a force of planetary cephalization. Tempering the mental features of this transition while remaining with the spirit of Grinspoon's argument, I would suggest the Sapiezoic eon marks a transition where the Earth is not only a living planet (with a biosphere at its disposal intra-acting with the habitable environment through biogeochemical processes), but *an artful planet* (with a technosphere at its disposal intra-acting with the habitable environment through *techno*biogeochemical processes). Such an artful planet would continue to stabilize the intra-action between habitable conditions and existent life breaking the cycle of mass extinctions—or at least significantly retarding the likelihood of such an event.

For Grinspoon, this is really the critical point of thinking about the Anthropocene in light of astrobiology. Astrobiology continually pulls the scale of our imagination outward. It continually presses us back to a planetary vantage point and a cosmic time-scale.[72] Doing this requires thinking about our impact, intentionality, and action in light of our proto-Anthropocene challenge at a *species-level*, despite individual and cultural differences. Certainly, Grinspoon is aware that we need to think about daily problems and cultural conflict alongside the wide vista view of an astrobiological approach to our proto-Anthropocene problems. He is clear we need both the quotidian and the cosmic.

Nonetheless, we see a strong resistance in the position he is developing to reducing the significance of the Anthropocene to just its role in constructing narratives about the relationship between "human being" and "nature." The Anthropocene may do this but it is simultaneously doing something bigger. The disorientation it effects is not only personal but also cosmic. There is a drive in his thinking toward acknowledging the intrinsic importance of taking this broader, cosmic view that he sees all too easily neglected in thinking about the significance of the Anthropocene in terms of narratives that so dwarf the historical perspective of human civilization that the Earth itself becomes the subject of our thinking—particularly in comparison to other planetary histories.[73] At present, human being is flexing the muscle of its geological agency, but what is important about the Anthropocene is to consider how it might induce a new eon in the history and development of the Earth: making a living planet into an artful planet—a task that may be carried out by a species other than *homo sapiens*.

## Anthropocene Disorientations

Before beginning to weave the threads of astrobiology, the *imago Dei*, and the Anthropocene back together in the penultimate chapter, we should take a moment to clarify how the Anthropocene is disorienting. The disorientation generates a sense of anxiety and provokes us to assessing our existential situation: clarifying how our encounters with ultimacy and the hope they imagine for our existence are not wholly realized. This generates an important parallel to the significance of the image/likeness distinction that formed part of the analysis of the *imago Dei* in Chapter 4: I am not yet what I am supposed to be. The Anthropocene, however, locates the concern for such disjunction in terms of the intra-active relationship of human being and nature: We are not yet what we are supposed to be. The anxiety felt over this unrealized expectation—over the gap between "is" and "ought"—is (at least potentially) a productive disorientation if that anxiety does not cripple us to inaction.

Briefly looking to the history of environmental reflection on nature, however, indicates that this concern for understanding the intra-action of human being and nature is hardly new. There is a robust and variegated history of conceptualizing environmental reflexivity that must be brought to bear in interpreting the meaning of the Anthropocene. Perhaps one of the greatest concerns the Anthropocene must face is to try and better understand why nature is consistently marginalized and backgrounded in light of the process of human meaning making despite the presence of a robust history of environmental reflexivity. Ultimately, the hope would be that, by gaining such an understanding, we might generate an alternative Anthropocene imaginary, based in a different set of categories, that would prevent the backgrounding of nature and preserve the prevalence of intra-action.

Planetarity and deep time were the proposed sets of categories that might do this work, though it is in their dialectical relation that they are most effective by preventing a recourse to self-contradictory tendencies in each category. In the Anthropocene, human beings are planetary creatures situated in deep time. Planetarity indicated our fundamental grounding in alterity: to be a planetary creature is to be under the species of alterity. This alterity cannot be reduced to our own subjectivity, but instead indicates a fundamental relationality that exists in the meaning of human being. We are embedded in and dependent on the web of planetary relations: Human being must be understood as part of planetary intra-action. Still, this embeddedness and dependency puts responsibility at the heart of a human way of being in the world.

Being situated in deep time de-centers any remnant of human exceptionalism. The temporality of human social history cannot be made separate from the temporality of Earth's geological history in the Anthropocene. The effect is that the scale of geological history eclipses the distinction between a backgrounded nature and meaning making of human existence. Human being is one of many participants with geological agency in the eons of the Earth; so too, it is only one of many species whose flourishing must be considered in imagining how to act out our right to be responsible.

In dialectical relation, the de-centering of human being in deep time tempers the risk of planetarity's responsibility becoming individualized (cutting against its very grounding in alterity). Geological

agency affecting Earth's history is never the property of an individual, but always of wider living-systems and populations. The responsibility of planetarity—pulled ever outward by the contextual
scale of deep time—is a distributed and shared responsibility that
deepens the sense of dependence and embeddedness that are part of
planetarity's alterity.

Vice versa, the alterity of planetarity tempers the risk of understanding deep time in terms of a technophilia promulgating human
exceptionalism (cutting against the de-centering affected by being
situated in deep time). Even if the Anthropocene is representative of
a new sort of technological catastrophic change (wherein the technosphere is integrated to existing biogeochemical pathways in the
Earth system—making a living planet into an artful planet), inaugu-
rating not only a new epoch with the "mature" Anthropocene but
a new Sapiezoic eon (as seems to be Grinspoon's hope), planetarity
consistently draws back any tendency to isolate human being or human capacities as *the* source for this change in geohistory.

If we understand ourselves to be planetary creatures situated in
deep time, we can disorient the received normativity and hegemony
of an everyday understanding of nature as that realm which is without human interference and the predominance placed on individuality in giving an account of human agency. These alternative categories aid the interpretation of the meaning of human being within the
Anthropocene and better enable us to cultivate "planetary knowledge and planetary identity,"[74] which gives credence to the intra-
action that has been stressed throughout this volume.

Effecting changes that respect these two sorts of disorientation
and take seriously how the categories of planetarity and deep time
help us interpret the Anthropocene by opening a new angle on our
way of being in the world is no small matter. To adopt these categories requires much more than merely being aware of them. It is not
as though we simply gain knowledge of planetarity or deep time and
find ourselves suddenly rejecting a lifetime of choices and values
that have been arbitrated under the assumed legitimacy of liberal
individualism. Knowledge is not enough. In the face of disorienting
knowledge, we need not necessarily effect a change in our way of
being in the world; we might instead delegitimize the disorienting
experience. What more than knowledge might be needed is a central
issue for the final chapter.

First, we need to weave together the three themes laid out so far
in order to imagine how this might make a difference to our way of

being in the world: making explicit what might be the implications of corroboratively interpreting these symbols of the *imago Dei* and the Anthropocene in light of astrobiology as our context of engagement. What does it mean to be the image of God when we take seriously the intra-action between living-systems and their habitable environments? And, how can this matter in an age of environmental crises where our well-honed senses of autonomy and individual agency have become disorienting?

# CHAPTER

# 7

# An Artful Planet

As part of our college's First Year Experience, each student has been reading Annie Leonard's *The Story of Stuff*. Leonard makes the case that—in light of the environmental crises we face and the finitude of planetary resources—we must change the relationship we have with the things we own: We must resist a disposable culture. She presents a panoply of digestible facts and figures meant to persuade general readers that there are significant problems with our current rates of human consumption and resource use.

Around halfway through the study of the text, an increasingly disgruntled student in my section said, "I'm sick of reading about the problem. When are we going to get some solutions? We know all these things are problems. I didn't know quite how bad they were, but I thought this book would help me think about how to fix them."

Another student pointed out that at the end of most chapters, and certainly at the very end of the text, Leonard provides suggestions about how we might change our behavior and think differently about the "stuff" we have and use. After naming a few of Leonard's suggestions, he closed his response by saying, "If more people knew how bad things really are, they would change the way they use their 'stuff.'"

"Those aren't solutions . . . that's just philosophy (no offense Dr. Pryor)," said the disgruntled student as she jumped back into the conversation, recognizing her slight at my department. "But, what I mean is that we can't rely on people changing their minds; we need real, specific steps to solve these problems. If we have to rely on people changing their minds and thinking about how they use and buy their stuff differently, this is a waste of time—it will never work. These problems are too big and just because people know about these environmental problems doesn't mean they will do anything about it. Even if they did, what one person does won't make a difference with these problems. We might as well just enjoy the time we have left as human beings because we are all going to go extinct."

I believe this sort of exchange has become more common in our wider public discourse. Moreover, as I scanned the room, I was a bit surprised at the number of students who were nodding in agreement with the disgruntled student as she spoke. There were clearly aspects of what she said that other students had been feeling.

This whole interaction was delightful for me as the course instructor. It gave me plenty of fodder for the next few class sessions. However, as a human being, I found the despair evidenced in my disgruntled student's answer and its resonance amidst the room full of eighteen-year-old students a bit disconcerting. The boundless optimism of college that encourages them to live into their vocational aspirations and change the world around them was already fading in the face of a despair over seemingly intractable planetary problems that minimize the agency of individual actors. Moreover, there was a deep sense of frustration at a perceived lack of reasonable solutions; despite the urgency attached to the problems we face, the students' demands for pragmatic responses was going unfulfilled. We might even more broadly claim that my disgruntled student was expressing a certain sense of a wider public fatigue with understanding the depth of the crises and anxieties we face today, while simultaneously feeling disempowered from meaningfully addressing them in any way.

What I stressed in Chapter 5 from recounting ways of thematizing nature is that we have a long history of environmental imagination shaped in the face of knowledge about environmental crises and degradation that has been ignored. Moreover, in Chapter 6, I contended in chorus with a host of others that the Anthropocene disorients our way of engaging with the wider habitable environment. If we take the Anthropocene seriously as a symbol for the intra-action of human beings and the wider habitable environment, then we must take seriously how we are situated both spatially in our planetarity and temporally in the deep time of Earth's history. Yet, if the sentiment of my students reflects a wider despair (as I suspect it does) and we have evidence that our societies have already ignored the voices of environmental consciousness in previous generations, simply introducing such categories for reconceptualizing our way of being in the world is not enough.

Resonances have already been appearing across the themes of the first six chapters, and I have already begun to draw some initial connections between those themes at various points. Now, we must begin braiding these strands of thought more tightly together. My hope is that in the intersection of astrobiology, the *imago Dei*, and the Anthropocene, we might discover a different way forward for addressing our impending environmental crises that does not rely solely on a logic of sufficient knowledge changing the actions of individuals. Specifically, I suggest the inclusion of the *imago Dei* in this intersection of themes is not merely adiaphora; instead, this doctrinal symbol may hold crucial significance for wider public intellectual engagement with the implications of our astrobiological awareness, the disorientations posed by the Anthropocene as a symbol, and the impending environmental crises we face on our planet.

## Doctrine Beyond Apologetics

Before moving directly to stitching together the three themes addressed in the book, I want to return to a methodological claim first offered in the introduction. Methodologically, this work represents an effort to move away from a traditional, apologetic approach to doctrine and instead think about doctrine as a mode of symbolic engagement. Symbols provide a connection to ultimacy—a way of our participating in ultimacy that summons us to live humanly together in the world as we adopt transreligious, public values. In this project, the *imago Dei* has served as a prototypical example of a doctrine taken up in terms of symbolic engagement. It is a helpful place to begin because it is so immediately applicable to how we envision what it means to "live humanly."

In moving away from an apologetic approach to doctrine and toward modes of symbolic engagement, the symbols, or what they inspire, are never significant merely *within* the originating community. The claims to meaningful being and the values then associated with that way of being have a public dimension. They express an aspirational value toward fulfillment and flourishing that extends well beyond the originating context and first interpreting community (even if such a history can never be left behind if a symbol is to continue to provide a meaningful lens onto ultimacy).

Apologetically, one might call what follows in this chapter the culminating summary of the book insofar as it functions as a constructive Christian outline for what it means to be the image of God when that term applies to the whole living-system of human being and not simply individuals. Speaking to confessional communities that already accept the doctrine of the *imago Dei*, this project can be taken as an extended consideration of how we might live-with and live-toward others into being the *imago Dei* during the Anthropocene when we recognize our place in the cosmos astrobiologically. It correlates a doctrine to a shift in the wider existential context of people who already accept the efficacy of that doctrine.

While this certainly could be helpful in specific communities, I hope simultaneously to be making a claim about a broader public theology and the value that doctrinal symbols may have in public intellectual discourse. To that end, the more traditional apologetic possibilities opened by reinterpreting the *imago Dei* as a sort of refraction outlined in Chapter 4—that might be most self-evidently significant—need not be diminished or contrasted with the wider public significance being envisioned.[1] After all, to enact this vision of being, the *imago Dei* requires participation and cooperation with others that extend well beyond any confessional boundaries. Even if *I* live fruitfully as an individual who participates in the process of refracting the creative power of the divine, *we* might fail to be the image of God. Moreover, even if my denomination or religion lives fruitfully into refracting the creative power of the divine, *we*—as human beings—may fail to be the image of God. Engaging these issues is not optional on the model of the *imago Dei* proposed here.

Two interrelated issues are arising and need to be addressed if my use of the *imago Dei* is not solely intended to be apologetic in function: a specific issue regarding the *imago* Dei and a more general methodological issue regarding how we understand doctrine to function as a symbol arise at this point. The specific issue is that if I am not religious or do not accept the traditional tenets associated with the *imago Dei*, then why should reflection on being the image of God be significant to me at all? The more general issue that follows from this is to consider more precisely how *any* doctrine taken as a symbol might foster a meaningful connection with ultimacy beyond the originating community of that symbol.

My focus is on the more specific of these two registers, dealing with the significance of the *imago Dei* for public intellectual discourse regarding our environmental imagination in light of being interpreted in an astrobiological context of engagement and as corroborative with the Anthropocene. However, before diving into the specific nuances of those interactions, I want to consider briefly the more general register: dealing with the potential of doctrinal symbols to function in public intellectual discourse writ large. Doing so should highlight how my consideration of the *imago Dei* might be an example of a broader way of methodologically reframing the use of doctrinal symbols for public discourse. In short, I want to briefly consider if there is a difference between the specific ways that doctrinal symbols might operate in public discourse from symbols more generally understood in order to posit that my more specific analysis of the *imago Dei* is an exemplar of this more general approach to doing public theology.

To think about the importance of doctrinal symbols, I want to return to a distinction introduced in Chapter 3 and specified in Chapter 5. Therein I highlighted that all symbols address an inherent tension of ultimate concern. The risk is that when a symbol tends toward expressing ultimacy, it can become meaningless through generalization, whereas if a symbol tends toward expressing our existential concerns in concreteness, it risks becoming myopic. Corroborative symbols become a check and balance on these tendencies to either meaningless abstraction or unhelpful concreteness. Specifically, I went on to suggest that the *imago Dei* was a symbol that emphasized ultimacy and could be corroboratively related to the Anthropocene as a symbol that emphasized the concreteness of our planetarity when interpreted in light of astrobiology.

I would suggest this is a generalizable pattern. Various doctrinal symbols—especially ones that are public-facing or that are engaged such that living-into the hope they promulgate requires participation with others beyond the confessional community—could be used in corroboration with other symbols that drive us toward a deepened awareness of the inescapable concreteness of our existential concerns from which we can meaningfully order our existence toward ultimacy. Doctrinal symbols, taken this way, would be symbols of hope that give us an instinct for moral revival and the formation of transreligious values concomitant to those symbols that clarify the contours of our context of engagement and make us more deeply aware of the anxieties that shape how we might meaningfully order our existence.

To this end, the concept of corroboration is crucial. In Chapter 5, I already noted that corroborative symbols mitigate the potential weaknesses of one another. Symbols indexed to ultimacy (as with doctrinal symbols) risk becoming meaningless through generalization when unmoored from the concrete concerns of a given place and time; vice versa, symbols indexed to our existential concerns risk becoming meaningless through a misappropriated concreteness. Symbols indexed to ultimacy need corroborative symbols indexed to our existential concerns. Without this corroboration, these symbols for ultimacy lose their power meaningfully to reorder our existence, which we experience in all the concreteness of its quotidian features. However, in corroboration, these symbols for ultimacy remain vitally important.

Reclaiming doctrinal symbols in public intellectual discourse, then, could be a powerful tool too often cast aside. We too easily assume that doctrinal symbols have died and are no longer relevant to the concreteness of our modern concerns and anxieties. This would, in fact, be true; but, to cast them aside for this reason is to misunderstand how doctrinal symbols might function in wider corroborative networks. The strength of doctrinal symbols is in speaking to ultimacy, not concreteness. When held in a corroborative tension, they can have a powerful influence in our public discourse; however, this is possible *only* if they are deployed carefully in wider public intellectual discourse.

If the doctrinal symbol reifies into a concrete theological answer that responds unilaterally to the existential questions of any time made concrete by other corroborative symbols, this is not helpful. The doctrinal symbol, in such a case, becomes hegemonic; it no longer works in corroboration to the symbols clarifying the concreteness of our existential concerns. Instead, in the face of the knowledge imbued upon its corroborative symbol of concreteness, a doctrinal symbol must grant courage to face the depths of our anxiety. Doctrinal symbols should be used in public discourse insofar as they can steady us in the face of this dizzying descent into anxiety that risks becoming utter despair.

It is worth dwelling on this point a moment longer. What I am suggesting is that we need symbols indexed to our existential concerns to help make sense of our anxiety. These symbols convert our anxieties into fears that are increasingly concrete. These fears can be grappled with, but they simultaneously increase our anxiety insofar as they clarify and specify it. Without some symbol indexed to ultimacy that gives us courage to respond, this cycle of anxiety concretized to fear through symbols indexed to our existential concerns leads us into a paralyzing sense of despair. A symbol corroborative with ultimacy breaks this tendency to enter despair by promising a mode of response-ability to those concrete fears we must face—a courage despite fear.

This is where the corroboration of symbols is crucial. If a symbol indexed to ultimacy only speaks to our sense of anxiety (never engaging the symbols indexed to our existential concerns that concretize this anxiety into fear), then it remains abstract and meaninglessly general. It is only in corroboration that a symbol indexed to ultimacy—such as doctrinal symbols—can become meaningful to various publics. Doctrinal symbols, in their connection to ultimacy and the promise of ultimate fulfillment, provide a courage to confront the disorientation induced by symbols that clarify the concreteness of our situation and orient ourselves anew in the midst of the world. If doctrinal symbols have this power, then they have a role to play in public intellectual discourse. They would be crucial to preventing a turn to despair in the face of the anxiety induced by symbols that primarily speak to concreteness and disorient our everyday way of meaningfully ordering our being in the world.

We might think back for a moment to my disgruntled student from the introduction to this chapter. Could a doctrinal symbol help her cope with the despair felt in the face of impending environmental crises? Does a doctrinal symbol speak to us in a way that prevents the intellectualist recourse to change she critiques? Specifically, do doctrinal symbols help us address the assumption that if we simply know enough about a problem that confronts us (or, in terms of the language I have been using here, as having only symbols clarifying the concreteness of our situation and anxieties) then we will simply make a rational change to our actions and behaviors that addresses this problem?

A doctrinal symbol, like the refractive account of the *imago Dei* offered here, should prevent the nihilistic tendency to submit to despair that tinged the end of my disgruntled student's reflection. We might analyze her interaction with the other student in the class as follows. Initially, she was frustrated with a lack of symbols indexed to ultimacy that sufficiently address the suite of symbols indexed to existential concern that were deepening her sense of just how direly serious are the concrete fears we face today. Using the *imago Dei* and the Anthropocene as examples, we could say that she felt how the Anthropocene was deepening her fears over patterns of human consumption and resource use, but felt paralyzed by this deepened awareness. She did not have a sufficient sense of symbols indexed to ultimacy to respond to this sense of increased anxiety that came with the specificity of her fears. The student suggesting that if more people knew the extent of these problems we could see positive changes in people's behavior is, in a sense, proposing that human rationality is a symbol indexed to ultimacy that can sufficiently respond to the increased anxiety she was feeling.

Clearly, however, it was not sufficient, and I do not think this should be surprising. In the student who appealed to human rationality and knowledge as a cure for environmental degradation, the symbol indexed to existential concern is made to do double duty. It is then meant not only to clarify our anxiety, but it assumes this clarification also alleviates the anxiety. By no means should we assume this is always—or even often—the case. The clarification the symbol allows in connecting concrete instances of fear to the more general sense of anxiety can deepen the sense of anxiety—even to the point of despair. Instead, a symbol indexed to ultimacy has to reframe the meaningfulness threatened in the state of anxiety in such a way that it addresses or overcomes the concretized fears.

My disgruntled student experienced this. Thus, in the face of this suggestion that knowledge is enough, she responded with a hedonistic call to dwell in the despair of our situation. A doctrinal symbol, like the *imago Dei*, though, might speak to her anxiety tending to despair in a different way than the appeal to human rationality allowed. It inculcates a *hope* that gives content to an alternative form of meaningful existence. This meaningful existence is not predicated on the same set of symbols that helped correlate our anxieties to more specific concrete fears. Instead, the doctrinal symbol that is corroborative to a symbol of existential concern addresses the concrete fears that are being manifest by subverting assumptions at work in our general anxiety. This can disrupt a slide into despair by providing a courageous way forward for shaping new transreligious values that speak to the problematic assumptions shaping our more general sense of anxiety.

My hunch throughout this work has been that doctrinal symbols speak to ultimacy. In so doing, they drive us forward in hope to meaningfully reorient our way of being in the world in light of how other symbols that speak to the concreteness of our existential situation disorient our everyday sensibilities concerning a given context of engagement for interpreting these sets of corroborative symbols. If this is the case, we should now be able to weave together our themes of astrobiology, the *imago Dei*, and the Anthropocene to see such a pattern begin to play itself out. It is to this more specific task that we must now turn.

## An Artful Planet

What I have already stressed is that we must interpret symbols in light of a given context of engagement that provides the hermeneutical framework in which the symbol can meaningfully inspire ways of being in the world and the formation of values in a given community. In Chapters 1 and 2, I considered how astrobiology serves as a context of engagement. Immediately, it became clear that astrobiology provides a wide context—one that, in principle, could be significant to any form of planetary life across the history of the universe!

This nearly unfathomable scope reflects the breadth of astrobiology's concern: considering the intra-action of living-systems and the wider habitable environment, particularly in light of how such intra-action might arise. This idea of intra-action is fundamental for astrobiology. If taken seriously, astrobiology assures that we cannot sufficiently understand either living-systems or a habitable environment in isolation. The ways of framing this intra-action are diverse; exoplanets, icy moons, and Mars all present different sorts of challenges to astrobiologists, and the intra-action that we imagine is particular to each case. Yet astrobiology gives us an important glimpse into the deep history of living-systems and habitability (in a way concomitant to the symbol of the Anthropocene) such that the ready distinction between these concepts—living-being and world—is made fuzzy and indeterminate. Adopting the language of planetarity, we might say the intra-action of living-systems and the habitable environment form a non-separable difference stemming from this deep history of their co-emergence.

Certainly, intra-action is a key to understanding the significance of astrobiology forming the context of engagement for interpreting symbols. However, the vastness of the cosmos that represents the scope of astrobiology is no less significant as it drives us back to the particularity of our location here on Earth and its importance to us as a living-system. A robust and serious consideration of the intra-action astrobiology considers forces us to question the legitimacy of cosmic escapist fantasies. We are integrated to the planetary flows of the Earth. We must go to great extremes (as with the spacesuit described in the introduction) to escape our tethering to this particular location in the cosmos.

This is significant in two senses that we should clarify. First, given our intra-action with this specific habitable environment across the vastness of the cosmos, we should be taking more care with our habitable environment. To be human is to intra-act with the planetary flows and systems of the Earth that are not easily replicated. Astrobiology should make us more deeply aware that to put these flows out of order, to generate runaway positive feedback in our planetary cycles, risks human extinction. The planet will be fine—it will move along with its biogeochemical cycles and recover over the course of deep time. Homo sapiens would simply be an aberrant blip in the cycles of this living planet—something that should, nonetheless, hopefully matter to anyone reading this book.

Second, an astrobiological awareness should drive us back to a deep appreciation for the peculiarity of place in another sense: We exist as participants in a specific habitable environment. In an astrobiological context that stresses the intra-action of living-systems and the wider habitable environment, would a homo sapien living on another world still be human? It is an interesting question, which science fiction authors have long explored. I think if we take the intra-action of astrobiology seriously, then to be human requires a tie to the particularity of our place in the vastness of the cosmos. To be a human is to be an Earthling. Even if homo sapiens did live elsewhere in the cosmos, their intra-action with another habitable environment would make them distinct from us here on Earth.

In sum, astrobiology makes us aware of our context as intra-active earthlings in the vastness of the cosmos. In an astrobiological context, we do not think, specifically, of individual living things, but the intra-actions that integrate entire living-systems, and even sets of living-systems, into the wider geochemical planetary flows of a particular habitable environment. Finally, this intra-action is considered in different astrobiological phenomena at different levels of scale (two of which, instantaneous and continuous planetary habitability, we identified specifically).

Having recapped the implications of taking astrobiology as a particular context of engagement, we can return to how the *imago Dei* and the Anthropocene as a suite of symbols corroboratively work to meaningfully order our existence toward ultimacy in light of the concerns of our existential situation. Looking back to the history of the *imago Dei*, I stressed that we cannot lose sight of how our existential experience is one of "not yet." We are not yet what we are supposed to be. No account of the *imago Dei* can gloss over this fundamental disjunction in our existential situation. The *imago Dei* is something we live-into not something we statically are.

To retrieve this dynamic dimension of the *imago Dei*, we need to do two things. First, we should reclaim the place of the *imago Dei* within the wider context of cosmogonies that illustrate the creative power of God to fashion meaningful existence, not merely some banal materiality. Second, we must acknowledge how the metaphor of reflection, so often invoked as the meaning of *imago*, is insufficient. Merging these ideas, we can imagine human being as a *symbol* for this divine creative power. Invoking this idea of symbol, I suggested we should understand image to connote more than reflecting (letting the creative power of the divine rebound off us as light reflects against a mirror): It is a refraction. As the *imago Dei*, we refract the creative power of the divine in existence, bending and changing this power so that new ways of meaningful existence become available to creation. We are the curious creatures that find our sense of meaningful existence in opening up new possibilities of meaningful existence through the flourishing of the rest of creation.

What the astrobiological context of engagement indicated and corroborative association with the Anthropocene further confirmed, though, is a need to resist individualizing interpretations of this refractive *imago Dei*. Historically associated with human distinctiveness, the *imago Dei* has traditionally related to features or properties possessed by individual human beings that set us apart from the rest of creation. However, in the context of astrobiology, we are fundamentally understood as part of wider living-systems and human being itself is understood as being a specific sort of living-system; moreover the Anthropocene indicates that the agency of individual action appears strikingly insufficient in light of concern for current environmental crises.

In the astrobiological context and in corroboration with the Anthropocene, we cannot interpret the *imago Dei* as if it is something any one individual has; instead, it is something that we live-into together as human being. In a curious sense, here I become my sibling's keeper. Neither of us can be the *imago Dei* alone. Practically, the *imago Dei* is a doctrine that is always looking outward to wider publics when interpreted in light of astrobiology. It describes a particular type of intra-action between a living-system and the wider habitable environment: one in which new possibilities of meaningful existence and the flourishing of creation are made possible. We bend the creative power of the divine through our intra-actions with the wider habitable environment so as to open new possibilities for all sorts of creaturely (not just human) flourishing through our presence. As an intra-action, living-into this vision of the *imago Dei* is never something we can do alone.

Taking this doctrinal symbol seriously, as it has been rendered here, it becomes incumbent upon human beings to develop public, transreligious values that facilitate the development of a flourishing environment and a more just social order. However we think about living out the *imago Dei* on this view, we must carefully consider what it means to live into a *shared* sensibility of this doctrine given its historical tendency toward individualizing. How do we participate together (in all aspects of our creatureliness) in refracting the creative power of the divine toward the meaningful existence and flourishing of all creation?

It is important to pause at this point and take note of one more important feature of interpreting the *imago Dei* in light of astrobiology: the scale at which we consider what it means to live-into the image of God is inherently flexible. I noted in chapter two there are at least two scales on which astrobiologists conceptualize habitability: instantaneous habitability and continuous planetary habitability. I suggest we think of the *imago Dei* in a similar vein—refracting the creative power of the divine in smaller groups on the scale of instantaneous habitability or as a species at the scale of continuous planetary habitability. We might even call these an instantaneous *imago Dei* and a continuous planetary *imago Dei*.

Simply put, we should not assume that thinking about the *imago Dei* in light of astrobiology means we ignore local instances of the emergence of this sort of intra-action. We need to celebrate the ways in which we can live into the *imago Dei* at different spatial and temporal scales. Rural communities, neighborhoods, towns, cities, states, or individual countries may all have ways that they can facilitate living-toward the *imago Dei*. In fact, almost any local environmental initiative we can imagine could serve as an example here: refracting the creative power of the divine in an instantaneous way to adopt a new sort of intra-action within a specific ecological system. These initiatives would all represent an instantaneous *imago Dei*. As will be discussed later in terms of facilitating wonder and play, these local and regional efforts are at least ways of imagining alternative intra-actions with the wider environment that might be scaled-up in the future to have wider planetary impacts.

Yet the astrobiological perspective should also remind us that focusing on these instantaneous intra-actions is not enough; we must also consider a continuous planetary *imago Dei*. This is particularly the case on Earth, where various living-systems have already fundamentally inserted themselves into planetary scale systems and we have a robust network of biogeochemical systems in place. Making the *imago Dei* corroborative with the Anthropocene prevents any simple recourse to the sufficiency of an instantaneous *imago Dei*. The planetarity and deep time of the Anthropocene disorient our tendency to focus merely on the local.

Planetarity, in particular, drives us toward a human responsibility as a way of being in the world that is always beyond the personal, respecting our status as being under the "species of alterity." My responsibility relies on the responsibility of my neighbor. In planetarity, neither of us can be human without the mutuality of shared responsibility. Moreover, planetarity pushes us toward pursuing responsibility that recognizes a flourishing that goes beyond the personal and even the inter-human to recognize our flourishing as one of many alterior aspects that participate in the wider networks of planetary biogeochemical systems.

There is a de-centering of human being in this view of planetarity made more complete in the cosmic perspective of deep time. The Anthropocene challenges any distortion that would divide the history of human being from the history of nature as though they were separate categories. Like the cosmogonic location of the *imago Dei*, the Anthropocene refuses to locate human being anywhere besides the wider, slower time of nature. In this way, deep time gives perspective to the extent of what planetarity's alterity implies. As Grinspoon succinctly described this, the deep time of astrobiology makes us rethink how to apply attributions more typically associated with species or individuals. For instance, in Grinspoon's work, "living" is often taken less as a descriptor for particular types of individual beings and more as a descriptive state for particular systems (biogeochemical systems) enacted on a planet.

A vantage point is opening on how we can describe an astrobiological *imago Dei* for the Anthropocene. In an astrobiological context of engagement we must interpret the meaning of the *imago Dei* in terms of being a specific sort of living-system intra-acting with the wider habitable environment. The *imago Dei* describes a particular type of intra-action, and this decenters humankind from the doctrine. We, as human beings, are not the *imago Dei* because we possess some set of faculties, relational capacities, or power over nature as it stands apart from us. Certainly, we might enumerate any number of faculties, relational capacities, and powers that are required in order to facilitate the intra-action of the *imago Dei*; however, it is the occurrence and sustained process of such intra-action beyond the level of the individual that actually constitutes us as the image of God.

We have to understand the existential anxiety that is part of this approach, particularly in deference to the environmental crises triggering our awareness of planetarity and our location in deep time that are symbolized by the Anthropocene. In light of these disorientations of the Anthropocene, we have to interpret what constitutes meaningful human being in a new way. It is a planetary phenomenon. In parallel to thinking of living-systems intra-acting with habitable environments in astrobiology, Anthropocene "living" is a planetary phenomenon that de-centers human being as we recognize our fundamental alterity. In so doing, the Anthropocene makes this feature of astrobiology as the context of engagement more directly significant to our own existential situation. Human being is a specific type of living-system that must be understood in light of its intra-actions; we might even say these intra-actions are heightened to a sense of alterity.

Now, we must think of "*imago Dei*" (and human beings specifically living-into the *imago Dei*) as a planetary phenomenon. To do this would mean that the *imago Dei* applies not to individuals, clearly, but to a certain state of the dynamic intra-actions at work on the planet itself, as a whole. Understanding the *imago Dei* this way requires situating this doctrine in the wider cosmogonies where the concept arises and more clearly articulating how the concept marks a shift in the structure of creation. Succinctly, *to be the image of God describes not a single species or genus, but a new state in the ordering and meaningful existence of all creation.*

Any number of creatures or species could be the *imago Dei* on another world or could *yet* be the *imago Dei* on Earth using this definition. Thus, this approach would not run counter to a de-centering of human being that is central to the way I have framed the significance of the Anthropocene. The important de-centering of planetarity that places us, human beings, under the species of alterity should remain firmly in place. Moreover, it also fits with the astrobiological context of engagement because the *imago Dei* might emerge as a new state on many other planets or planetoids as well, so long as they are marked by a particular shift in the dynamic intra-actions of the living-systems and habitable environments.

If we locate the *imago Dei* in the wider account of cosmogony as the introduction of a new state of forming meaningful existence, then the term signals a transition in the state of creation itself. I would call this a shift from a "living planet" to an "artful planet" through the presence of a refracting agency: a causal power that emerges and bends the creative power of the divine to make possible otherwise impossible configurations of meaningful existence and flourishing of the whole panoply of creation.

Still, we must characterize this transition from "living" to "artful" in a more technically rigorous way if we want to describe the *imago Dei* in a thoroughgoing, intra-active fashion. (One could still interpret the "refracting agency" of the previous step in a way that breaks apart the intra-active phenomenon so that human being as a specific sort of living-system is understood as some sort of causal power utterly separable from its wider habitable environment.) The previous chapter suggests that this transition from "living" to "artful" is characterized by the emergence of *techno*biogeochemical systems. The introduction of technological catastrophic change and the development of a technosphere layered onto the geosphere and biosphere provide a more rigorous way of characterizing this new state of meaningful existence.

This represents a subtle but significant shift in perspective. Human being is not a separate agency technologically acting from outside otherwise biogeochemical cycles. Instead, the technological agency of human being is woven into existing cycles as a planetary phenomenon. We are not technological actors with the living planet's biogeochemical systems available to our disposal; we are instead part of an artful planet in which technobiogeochemical cycles are emerging.

Here, we might think back to Jeremy Davies's example, introduced in the previous chapter, of a modified carbon cycle that includes an economic phase. How many other biogeochemical cycles in the sciences could be modified to include the effects of technological catastrophic change? If we are to take seriously our recognition of this shift from being a living planet to an artful planet, we need to stop thinking of technological developments as emendations to otherwise pure, pristine, uninterrupted, or *natural* processes and cycles. The public presentation of various planetary flows and cycles needs to include technological phases.

Moreover, we need to start acting like an artful planet—as creatures inseparably intertwined to the wider history and cycles of the Earth.[2] I already emphasized that the *imago Dei* is something we are and we strive to live into; put into the language of intra-action, it is not a foregone conclusion that the introduction of technological catastrophic change will yield a greater sense of flourishing and possibilities of meaningful existence for all creation. More technically, this means focusing on *how opening up new possibilities of meaningful existence to creation as the* imago Dei *fosters techno-biogeochemical cycles not subject to runaway, positive feedback.* Specifically, to be the *imago Dei* is to be a planetary system that weathers catastrophic change without the cycle of mass extinction events typical of the Phanerozoic eon.

This is a truly transdisciplinary definition of the *imago Dei*: It cannot be contained by any single traditional disciplinary field. It relies on joining together insights from the humanities and natural sciences in order to offer a direction for future research by which we can live-into being the *imago Dei*. One can begin to imagine a whole series of scientific research projects that would facilitate the development of the *imago Dei* in terms of developing technologies that foster technobiogeochemical cycles not subject to runaway positive feedback.

As a philosophical theologian, my own interests more specifically lie in the dimensions this definition opens for continuing work in the humanities. Specifically, because a single individual or specific religious group can never enact these concepts sufficiently, we must think critically about how we foster a concern for living-with one another as an artful planet. This notion has two important features.

The first is that the definition offered does not place the responsibility for fostering these technobiogeochemical cycles equally on all human beings. We live-into this vision of the *imago Dei* together but we do not all hold the same access to technologies or are able to employ technologies (either directly or indirectly) in ways that affect the biogeochemical cycles of the Earth. How do we begin to distribute this sense of responsibility justly and hold ourselves as individuals accountable to a planetary responsibility that we always enact in mutuality?

Second, how do we continue to invite those for whom this suite of symbols and context of engagement do not immediately foster a connection to ultimacy to address the pressing issues of concern and anxiety that we face today? How do we invite them to participate in enacting technobiogeochemical systems toward planetary flourishing if they are unmoved by an account of the *imago Dei* or the Anthropocene? Engaging these questions is not optional on this model of the *imago Dei*.

While the first of the preceding questions is beyond the scope of my work here, the second question is crucial to any consideration of public theology. It will not be enough analytically to declare what the meaning of this suite of symbols, the *imago Dei* and the Anthropocene in astrobiological context, is in terms of refractive participation, planetarity, and deep time. We have to determine how we inspire an empathy, an *Eindeutung* and *Einfuhlung*,[3] so that the experience of refracting the divine creative power as planetary creatures situated in deep time is made available to those for whom these symbols have not previously been directly meaningful—have not conveyed a participation in ultimacy—and may never be directly meaningful. On this account, any individual's own experience of the fulfillment and meaning promised by participation in ultimacy is radically dependent on our living-together into being this refractive image of God. What, then, are the themes or moods we must adopt and foster in others in order that we can live out, and invite others to share in, the fulfillment promised by living into our intertwinement with the world? What qualities or values do we need to foster in the wider public to have an experience of being human that can move us toward an appreciation for the symbolic ideals of the *imago Dei* and the Anthropocene? How might these conflict with or challenge existing values that shape our sense of meaningful human being and its socio-political realization? Particularly, how does this shift in the scale of our thinking (conceiving the *imago Dei* at the level of being an artful planet instead of interpreting the symbol in light of traits specifically realizable by *individual* human beings) not become nihilistically alienating?

One might call what follows in the final chapter an effort to think through how we resist the tendency to make our neighbors (both human and non-human) "disposable."[4] This will require a realigning of certain values. We must resist tendencies toward privatization and independence becoming modicums for efficiency as a mode of personal convenience realized through the disposability of others.[5] But, in a global technocracy that makes the world into disposable, standing reserve meant to benefit the self, how do we reclaim the planetary intra-action where self and world are chiasmically inter-twined? How can those symbols that put us in touch with ultimacy help us refashion a vision for being an intra-active self: working personally to live humanly together and recognize how we enfold and unfold non-separable differences yielding shared agency and shared responsibility for our artful world?

CHAPTER

# 8

# Living-Into Presence, Wonder, and Play

In Ursula Le Guin's novel *The Dispossessed*, she recounts a scene in a special sort of pre-school where the children are gently corrected for their possessive tendencies: That's not your blanket, it is the blanket you are currently using. The scene occurs in her rich description of a society on a moon where ownership and possession are made to be morally reprehensible in contrast to the society on the neighboring planet where these capitalistic values matter most. In the society on the moon, Le Guin explores both the beauty and tremendous sacrifice entailed by living in a society where a fundamental sense of shared responsibility and accountability is essential to meaningful personhood.

We can imagine this world, in fiction, but in practice generating this sense of empathy, shared responsibility, and shared accountability—critical values for enacting the sense of *imago Dei* as a planetary phenomenon—is no easy task. As the close of the previous chapter suggested, though, we must consider this issue if we take seriously the definition of the *imago Dei* offered here. I have contended that knowledge is not enough, but how do we inspire a sense of felt empathy for the technobiogeochemical cycles of which we are a part? How do we inspire others to meaningfully re-orient their world around the disorientations of the Anthropocene and the fulfillment of opening new possibilities of meaningful existence to creation? And, why might such a process happen more readily for some people than others?

Here, I suggest three pivotal themes for affecting others such that they become more attentive to participating in the *imago Dei* (not

146

directly being it as an individual): presence, wonder, and play. The order that I present them in here is not meant to be serial or sequential. These themes are equiprimordial existential structures that aid us in ordering our human experiences. It is not as though one always achieves presence first and then pursues wonder and finally enacts play. If anything, it might be that, practically speaking, we reverse these: We play, which gives a proclivity for wonder that can be formalized into a disposition of presence.

Though equiprimordial, these three existential structures are intimately related but hardly equivalent. Presence, broadly speaking, is a disposition or orientation; it names an existential commitment to being aware of the intra-active, non-separable difference that subtends our lived experiences of being a self in the world so that we do not fall apart. It is a commitment to expose ourselves to others as part of the elemental flesh.

Wonder is a mood or an attitude. It allows the world to appear to us with an openness or "making proximate" of those desires that are askew to our predominant, norming orientations: an attentiveness to disorientations. It names the continuous need to be aware of how the orientation of presence might need to be redirected toward those desires: making what was once unreachable proximate in a new way of being-with while resisting the temptation to simply put these askew desires in line with our normative orientations.

Finally, play is an expression of freedom. It liberates us from ordinary life. Never forced upon us or necessitated by material interest or profit (it does not fulfill our immediate wants and needs as a means to some specific end), play allows us to explore and construct worlds of other possibilities that are engrossing. In play, the lines between the world of the ordinary and the worlds we pretend blur and break down; with these acts of imagination, we act "as if" this possible world *is*, which then imparts meaning to our ordinary living in the world. Play not only provides the freedom to imagine and engage with those newly proximate desires discovered in our wondering, but its blurring and breaking down the boundaries of what is pretend and what is real facilitates a proclivity for wondrous encounters.

In developing presence, fostering wonder, and practicing play, we may just start, perhaps without even realizing it, to live into the responsibility of being planetary creatures living in deep time that refract the creative power of the divine—finding new ways of meaningfully existing together as an artful planet.

## Presence

Presence indicates a certain participation in the flesh. Both the terms "presence" and "flesh" are philosophically loaded. I invoke the term "flesh" (or what has sometimes been called "flesh of the world") in the spirit that it has been used by Maurice Merleau-Ponty, particularly in his latter writings as his phenomenology tended toward an indirect or elemental ontology. In a single sentence, flesh describes how Being is always given over as a depth that is indirectly communicated by the mutually institutive intra-action of beings characterized in terms of the incarnating of the body and adventing of the world as it is intended to overcome a Cartesian divide between the self as a subjective consciousness (res cogitan) and the world (res extensa).[1] It designates an intertwining or enlacement—an Ineinander—by which self and world unfold together.[2] Though this is a slight shift in my philosophical language (from the more predominant use of non-separable difference or intra-action), it should quickly become clear how the flesh is congruent with these preceding ideas.

To understand the flesh, we must always remember that to be one who senses things in the midst of the world is simultaneously to be one who may be sensed. It is easy to forget about this basic aspect of our participation in the world; and, we can think of instances where we find ourselves jolted back to awareness of our participation in this carnal ensemble. Immersed in his drawing yesterday, my son started when I touched his shoulder. There used to be a window behind the desk where I sat in my office, and I was often startled when students knocked on it (to let me know they were coming in or to playfully watch me jump) while I was reading. Insofar as I have a place, being thrown into the world, I am made available to the world: I am inescapably compossible as part of the flesh.[3]

The flesh indicates carnality as characterized by our sharing in corporeality; we are a body in the midst of other bodies that comprise the flesh.[4] While it is tempting to think that this sharing in corporeality simply connotes that we are all material beings, aspects of the physical world made of common stuff, this would not be correct. "Flesh" indicates much more than some common atomic structure or existence as bare materiality; this is because flesh is not exactly "something" in this substantialist sense.[5]

Instead, the flesh is like an "element." The elemental typically designates something that cannot be well accounted for by traditional philosophy; here, the concept is invoked to draw a clear contrast between the flesh and any substantialist approach to ontology.[6] Flesh is, instead, "the formative medium of the object and the subject . . . one cannot say that it is *here* or *now* in the sense that objects are; and yet my vision does not soar over it . . ."[7] Or, as David Macauley describes it, flesh "serves as a 'concrete emblem' of a way of being."[8]

This element of flesh is always realized *indirectly* and manifest *between* me and the world synchronizing these disparate realities. Thus, to understand this sense of the flesh as element, we have to consider how it is manifest through the encounter of beings because I cannot engage the flesh except through these encounters, even if it is not sufficiently characterized by any one of these particular encounters itself. As Glen Mazis rightly suggests, flesh stands between the general and the particular: a mediate—between the Being of ontology to which we are given indirect access through its happening in the contact of particular beings in the sensible world.[9] Perhaps it is best to say we catch glimpses of the flesh: not as a "something" we can directly set ourselves toward, but as the prereflective, abyssal grounds or insights that are realized through, *but also make possible*, the particular, specific sensible contact of beings.

An example can be helpful. However, in considering any example we must be attentive to how glimpses of this flesh emerge in recounting meaningful experiences where the particular constellation of the encounter opens new horizons for experiencing the world or working in it. Moreover, we must remember these encounters can be common. They are not only moments of great profundity sensing an underpinning harmony to the world (*kairos* moments of Christian theology) that occur infrequently; rather, there is a banal quality to the flesh such that, with a certain sense of attention, we might find it in *any* set of encounters.[10] With these caveats in mind, I will use an example that is something I do everyday—riding my bicycle.[11]

I rode my bicycle to work in my office this morning. I grabbed the dropped handlebar of my road bike and felt the gritty texture of the handlebar tape under my palm. As I swung my leg over the top tube, my knee bumped into the firm nose of the saddle sliding the whole frame a bit underneath my body and my hand slipped onto the rubbery hood of the brake. In the space between my thumb and forefinger I could feel the pressure of my body leaning into the hood as I tried to regain my balance and keep from falling over.

In this initial aspect of the encounter—where I feel features of the bike—there is a chiasm at work. Chiasm refers to a crossing or intersecting in the field of the flesh. Anatomically, it refers to the crossing of nerves behind the eyes that make our binocular vision monocular.[12] Phenomenologically, the term "chiasm" preserves a certain sense of this anatomical crossing—where two things become one. More technically, the chiasm in a phenomenology of the flesh describes the encroaching of a perceptive event (for example, my hand feeling the handlebar tape, touching the brake hood, or my knee bumping the saddle).

The crossing of the chiasm does two things that we can immediately note. First, it makes me aware of the obverse quality of the body. Previously, I described the shared carnality of the flesh: The flesh occurs as bodies in the midst of the world that in sensing are able to be sensed. Here, this sensing and being sensed is expressed as a quality of my body enacting the flesh. The lived-bodies of the flesh exhibit a *reversibility* in themselves.

Second, the crossing or intertwining of the chiasm is evocative. It is not just that my hand and the handlebar tape happen upon each other. The encounter yields a grittiness. Similarly, the brake hood that my hand slips onto gives a rubbery sensation and the nose of the saddle offers up a resolute firmness eventually bruising my knee. These encounters of chiasmic entwining evoke a sense that neither I nor the bicycle possess beforehand; nor is this encounter well described by some combinatorial logic (the sensibility is not something additive). A distinct way of being or quality of the world is emerging through this crossing of the flesh; as we mutually inhere to one another, a different way of being in the world begins to take shape.

The experience of reversibility is never simultaneous.[13] I never sense my own sensing. As I touch the handlebar tape I sense it, but while regaining my balance I am not so much sensing the bike as being sensed; the bicycle pushes against me—pressured against my body—not yielding to its tipping over and the intentions of my own body.[14] It is crucial that this experience of sensing and being sensed never collapses into an undifferentiated unity (a sensing my own sensing) because the crossing of chiasm and the various senses of the flesh that different configurations of my encounter with the bicycle evoke would not be possible without a distance or separation. If I *am* the bicycle and the bicycle *is* me, there is not a sufficient sense of distinction to have a crossing. The chiasmic encounter of the flesh, to be sensed and sensing that keeps the perceiver and the perceived in the same order of beings, relies upon this difference: There remains a separation or a distance between us—an *écart* or dehinscence.

This separation or distance, though, is not a hard and fast concept sharply dividing perceiver from perceived in strict opposition. Such a concept would violate the carnality of the flesh and could never accommodate the evocative quality of lived-experience that the chiasm describes. Instead, the separation or distance is an opening (as is well-described by subsequent deconstructionists). It is an opening or a space between the perceiver and the perceived that makes any quality of sensibility possible. To better understand why this is not a sharp division, lets return to the encounter with the bicycle.

Having regained my composure (and giving a quick check to see if anyone has seen me nearly tumble while inflexibly getting on my bicycle), I push hard with my right foot onto the pedal. It creaks slightly against my weight, but the whole frame seems to leap forward and the saddle settles gently underneath me. I never cease to marvel at how every downward push on the pedal seems to throw me forward on this composite frame as opposed to the more mushy lurching of the first steel-framed road bike I ever owned. A few more turns and I feel the cleats on my shoes snap into the clipless pedals and my legs spin more easily and freely as the road opens onto a long downhill stretch. It is cold today and the wind bites at my cheeks as I shift down the rear cassette so that my legs do not start spinning too fast on the pedals during the descent.

As I ride down the hill, feeling the pedals turn over under my feet, any separateness of me and the bicycle gives way to our intertwining. We move around potholes opening up before us as my arms extend into the handlebars so that the bicycle is incorporated into the space of my body. My legs are not separate from the pedals anymore but become pistons that pivot on a confusion of feet and crank arms so that the bike leaps forward, as though it wants to and I am merely there to make this expression of itself possible. I experience the bicycle at different moments as separate and intertwined: distant and then inseparable from the lived-space of my body—me and then not-me.

This tension between the experience of entwining and dehinscence (the inhering penetration of the chiasm and the persistent separation of the *écart*) defies our typical understandings of self and world, but it arises by looking into the sensed depth of these evocative encounters that we are naming flesh. My riding the bicycle helps illustrate a swift movement beyond any imporous border of individuation—as though my sense of being remains strictly divided along a dermal barrier. In the intertwinings and gaps of perceptive events, there is an incarnating of the self. *I* come to be through my perceptive incorporation of the bicycle that is no longer separate from me. Grammar hardly seems to do justice to the experience this chiasmic encounter evokes: "We *is* a deft vehicle descending the hill." At the same time, there is an "adventing" of the world in this act of incorporation—a promise of manifesting otherwise silent possibilities. Slaloming pothole obstacles and the wind biting at my cheeks are aspects of the world that could easily go unnoticed without the speed that my entwined body-bike achieves going down the hill. Maurice Merleau-Ponty calls this entwined development "a *pregnancy* of possibles, *Weltmöglichkeit*."[15]

> One can say that we perceive the things themselves, that we are the world that thinks itself—or that the world is at the heart of our flesh. In any case, once a body-world relationship is recognized, there is a ramification of my body and a ramification of the world and a correspondence between its inside and my outside, between my inside and its outside.[16]

Yet, what of the *écart*? It cannot be left behind so that our chiasmic intertwining simply becomes simultaneous; However, is that not precisely what has happened in descending the hill? While we can easily flip back and forth between sensing and being sensed as distinguishable conditions, there is something that our perceptual experience shows to be artificial in this division. We lose something of the perceptual phenomenon if we adhere to a stark separation of divisible moments in our fleshly encounters.[17] These divisible moments encroach upon one another so that finally the flesh evokes something distinctive—the "we is" of the body-bike—that pushes the *écart* to a new frame of reference felt at the borders of the bike tires with the road or the tingling skin of my cheeks; these two experience are no longer so disparate as they once might have been understood to be, as if one belonged to the bicycle and the other to my body. The experience of me and the bicycle as not-me crossing in the flesh falls away here in our *availability* to one another. Instead, a newly incorporated, hybrid—even cyborg—body advents upon a world that is now felt differently; a new set of chiasmic crossings in the flesh becomes possible, adventing new ways of being in and understanding the world.

When the flesh, with its chiasmic encounters, becomes *the primordial unit of our ontological thinking*, we find ourselves to be in a state of flux. The shape these entwinings of the flesh may take are not permanent. After all, I eventually arrive at my office and hop off my bike. My feet and legs no longer turn as pistons over wheels that were the source of my felt contact with the ground. Now the weight of my body rests on my heels as I walk awkwardly over the cleats in my shoes and my feet give me a lived-experience of the ground once again. The mutability of my various intertwinings of the flesh becomes clear. The sensibility evoked is not permanent nor is it forever gone; the flesh is mutable—crossing and uncrossing, tangling and untangling—letting these instances of evocative encounter rise up and fall away.

This indirectly revealed flesh makes for promiscuous beings; beings constantly engaged in the persistent metamorphosis of incorporating and adventing made possible as instances of the flesh's non-collapsed intertwining. Such promiscuous beings, engaging the dynamism of the flesh, yield a poetic and fragile ontology. It is, to borrow the idea from Bachelard, a penumbral ontology wherein the borders of ontological units are fuzzy as they bubble up in the contexts of particular practices and modes of working in the world, only to be reshaped relationally under the guise of new contexts.[18]

Even without our knowing, the flesh can silently spin along: birthing its "pregnancy of possibles," manifesting its "*Weltmöglichkeit*" in tandem to a perceptive deafness on the part of the individual who chiasmically enacts configurations of the flesh without any awareness. This is easy to do. The flesh is only ever indirectly manifest through its thematization in the encounter *between* beings—between a self and a world. The relative solidity of these beings that we directly experience can easily be assumed to be the primary and primordial unit for ontological reflection, in which case the flesh remains a hidden depth.

For this project, letting the flesh remain hidden just will not do. Given the astrobiological context of engagement and the corroborative symbols developed here, I am not convinced the concept of the *imago Dei* can be pursued at all without a wider ontological refiguring like the flesh provides. A concern for the chiasmic intertwining of the flesh that de-centers our autonomous individual sense of self by making our evocative experiences primordial would be a precursor to living into an intra-active conception of the *imago Dei*. Understanding our individual experience in terms of the flesh becomes a propaedeutic for participating in the larger scale endeavor of the *imago Dei*, wherein human beings are, and hope to evermore become, a site for making new ways of meaningful existence available to creation, particularly in a way that respects our Anthropocene status as planetary creatures situated in deep time. If we cannot evoke a meaningful understanding of our own alterity that challenges our sense of autonomous individualism, then we have little hope of developing political values by which we can realize this species level ability to be the *imago Dei*. Without an alternative ontological unit, the assumed status of the autonomous individual will continue to hold sway; in its sway, it is to the autonomous individual that we are morally and politically responsible. Understanding ourselves in terms of the flesh begins to break the stranglehold of that sense of autonomy. If we are to claim the power of intra-action between living-systems and the habitable environment, if we are to resist the disposability of our current technocracies, if we are to enact the responsibility that is our right as planetary creatures, then we must be *present* to this subtending power of the flesh.

Presence indicates more than a mere awareness of the workings of the flesh. It implies a certain disposition toward the chiasmic intertwinings by which the flesh is creatively working itself out through the various encounters of beings in existence. In short, it is a way of *paying attention*: of being present toward our abyssal grounding that we participate in shaping through every one of our encounters with other beings in the world. It is a way of being in the world that looks to foster the dignity and flourishing of the flesh through the ways we engage in various chiasmic encounters that constitute our existence.

Doing this requires treating each moment I take up the flesh in chiasmic encounter as a potential "crossed phenomenon." Jean-Luc Marion uses this term to describe the experience of loving. I would suggest his phenomenological description of love can be read as overlapping with the account of the flesh that I have offered here.[19] It describes a specific sort of encountering that can go on in the flesh, or we could alternatively say it describes a specific way of entering into the chiasmic encounters of the flesh.

In the crossed phenomenon, we begin with a lover deciding to love.[20] Love is an exposure and an advance on such an account. As an exposure, love is a self-offering by which I vocatively make myself available to the beloved. I declare, "Here I am," and in so doing I define my own meaningful being by a relation of presence.[21] This presence is on the one hand a sort of availability or advance wherein there is a continuous and persistent assurance of this declaration.[22] On the other hand, it is a sort of absence; I am present by making myself—my very particular and unrepeatable place in the midst of the flesh—into a space wherein the meaningful being of the beloved can flourish and find fulfillment. There is no expectation that my love will be returned when offering this declaration, and because there is no expectation that my love will be returned, there is a risk and vulnerability in this vocative pronouncement. As such, the address is also a form of exposure.[23]

The exposure and the advance of love—making myself into a space where the beloved can flourish and find fulfillment—assures the meaning of my being. It ensures that my own existence is not in vain or useless because I am the alteriority in which the flourishing and fulfillment of the beloved is possible.[24] The crossing in this phenomenon occurs in that my act of love phenomenalizes the beloved as one who is *also* a potential lover. I phenomenalize the other as one who might make space for others to flourish. I can never confirm if this potentiality is realized; the love the beloved may offer in return always remains alterior to me.

In short, the love offered by the lover for the beloved as a signification of presence—a "Here I am" that makes me available to the flourishing of the beloved—is in itself not available to the beloved. The beloved does not feel my own act of exposure and advance as the lover. Vice versa, the love made by the beloved toward the lover would be its own distinct signification of presence alterior to the lover. I do not feel as the lover the exposure and advance of the beloved in her act of offering love.

Though there are two irreducible significations, the significations themselves are equivalent. As such, "I do not individualize myself by self-affirmation or—reflection, but by proxy—by the care that the other takes with me in affecting me and allowing me to be born of this very affect."[25] The lover and the beloved are mutually constituted by this exposure and advance of love to each other.

What I have called "presence" to the flesh mimics this exposure and advance that a lover offers to the beloved in the crossed phenomenon, if we imagine each chiasmic crossing of the flesh as a potentially crossed phenomenon. Our opportunities for chiasmic intertwining with other forms of existence in the world should be guided by the openness of a continual and persistent exposure in which my body is understood as a space in which those other aspects of existence find a way to flourish through inhering with me. My body is a space of readiness for discovering the dynamic potentialities that might unfold through intertwining. My persistent exposure to the world as such an availability given to the flourishing of other forms of existence signals my responsibility for discovering the evocative possibilities of the flesh that otherwise remain hidden. This responsibility is very much like the sense of "responsibility" that Gayatri Spivak cited in the previous chapter, described as a fundamental right of human being in its planetarity.

To summarize, the different constellations by which we become chiasmically intertwined with other beings in the flesh open new vistas on our way of being in the world. The flesh is mutable and dynamic; these different ways of taking up the flesh are not permanent, and some may turn out to be more preferable than others. However, maintaining a general openness to multiple ways of taking up the flesh ensures that we might discover how these new horizons open up meaningful ways of our being-together that might otherwise be utterly unanticipated. In "presence" we call out to the world "Here I am" and invite encounter to develop meaningful chiasms.[26] We form ourselves into pursuing a practice of exposure and advance toward the alterity of the world to find ways in which the flesh—the primordial ontological unit from which our sense of self can only be artificially removed—might be fulfilled and flourish.

## Wonder

The self-risk of exposure and the advance of our being present to the flesh are hardly typical ways of responding to the world. Even having a disposition to be aware of our encounters with things in the world as chiasmic intertwinings defies our everyday sensibilities. To develop this disposition of presence toward the flesh that makes us into spaces where new ways of flourishing with other beings may develop through our being enlaced together requires an openness in the face of the alterity of other beings existing in the world. How do we develop such openness? How do we foster an orientation toward fleshly flourishing such that we engage in the risk of exposure rather than shirking from chiasmic encounters?

Queer phenomenology can be a helpful resource in this regard. It begins by rethinking "the 'orientation' in 'sexual orientation.'" Sara

Ahmed, for instance, offers "a phenomenological approach to the very question of what it means to 'orientate' oneself sexually toward some others and not other others."[27] While we tend to immediately associate sexual orientation with identity (gay, lesbian, bisexual, straight, and so on), queer phenomenology encourages thinking about how being oriented has two distinct senses. In the first sense, an orientation means being directed *toward*. In this sense, an orientation is a desire for something we do not actually have. I am oriented-toward that which is separate or distinct from me—given that the distance from me cannot be so great that the desire becomes utterly unavailable.

Implied in orientation-toward is a consideration of proximity. We can only be oriented-toward those things that are *proximate* enough as to be available to our perceptive bodies. Proximity does not indicate mere distance or nearness but *availability*.[28] To be oriented-toward something is to indicate that the object of desire is available to me; the various orientations we take (or that we inherit) shape what sorts of objects, feelings, thoughts, or spaces become available or unavailable as such.

The concern for proximity is real if we keep in mind that we cannot take up limitless objects. We are finite, limited beings. Though a seemingly banal observation, to be a finite creature oriented-toward certain things simultaneously implies we are *not* oriented-toward some other set of things. Not all the sorts of objects, feelings, thoughts, or spaces we might be oriented-toward are, nor can be, equally proximate. To take up one possibility and orient-toward it is to let another possibility lie.

In the second sense, orientation connotes being *around*. Instead of *taking* up (as with orienting-toward various proximities), being oriented-around entails our *being taken* up. An example can be helpful to make the difference clear: "I might be orientated *around* writing, for instance, which will orientate me *toward* certain kinds of objects."[29] Our orientation-around guides what we are most likely to be oriented-toward: It centers and binds us in a trajectory.

In a move that should not be surprising for anyone familiar with queer theory or gender studies, this distinction between orienting-toward and orienting-around indicates "that bodies are shaped by what they tend toward, and that the repetition of that 'tending toward' produces tendencies. We can redescribe this process in the following terms: the repetition of the tending *toward* is what identity 'coheres' *around* (= tendencies)."[30] Performativity is at work here in shaping our orientations.[31] Our orientation-toward certain proximities develops a sense of being oriented-around that reinforces what becomes proximate and likely to be oriented-toward in the future. This circular process absconds itself, however, and seems to reify what we orient-around as "natural." Part of the reason for this reification is that none of us produces our orientation-toward certain proximities *ex nihilo*. We receive orientations-around as a sedimented history, which continues and propagates through our continued adoption of the orientation-toward certain favored proximities.[32]

When orientation is not simply a marker of identity, it has to be interpreted in the richer intercorporeal matrix of the *body, objects, and spaces*. Orientation, whether as orientation-toward or orientation-around, does not imply a subjective, willful power of consciousness. Nor is it a purely intersubjective phenomenon: "Bodies as well as objects take shape through being orientated toward each other, as an orientation that may be experienced as the co-habitation or sharing of space."[33] Orientation is a trajectory, a way we realize or expect to realize a meaningful formation of bodies, objects, and spaces together.

This sharing of space is resonant with the account of the flesh offered in the previous section.[34] The extension of the body into spaces and objects and the incorporation of objects and spaces into bodies as the chiasmic intertwining of flesh are formative processes that give rise to different senses of being a self in the midst of a world. The example of riding a bicycle illustrated this. The bike could be part of the world (placing a gap between my body and the bicycle as part of the horizon for my world) or part of my body (placing a gap between this different sort of embodiment—the bike-body—and an alternative horizon for my world). The flesh was the element underpinning these experiences. Never directly experienced, the flesh could be folded and refigured into a panoply of mutable shapes for my body and the horizon of the world to take. The diverse ways I could experience the intersection of my body with the horizon of the world was a product of the various ways that my place in the flesh became chiasmically entwined with other existent beings.

If orientations belong to neither the body nor objects and spaces alone, orientations also sit "between"—much like the flesh. I suggest it is the flesh, that ontological element always "between," that takes on orientations as a tending toward and developing tendencies around repeated chiasmic encounters. On such an account, orientations, too, would be indirect (we do not experience orientation as its own thing or space but only as a drawing toward or around some set of things directly experienced), but they would have significant effects on our direct experiences of being a self in the midst of a world.[35] Now we can also clearly see that such shaping of the flesh has consequences; taking up the flesh in orientation-toward a particular object or space makes other objects and spaces more proximate and others less so. How the flesh is oriented-toward would describe the possible ways the flesh could be folded and figured in chiasmic encounters to give different shapes to the body encountering the horizon of the world. How the flesh is oriented-around would describe a proclivity for certain figurings for the flesh shaping patterns the self and world might take, canalizing how the flesh might be oriented-toward particular sets of chiasmic encounters.

If orientations (whether -toward or -around) are indirect, we only experience these orientations insofar as they are manifest through direct experiences. They are experienced by proxy—a facet of our direct experiences that can remain hidden and silent without attention. In this way, orientations are much like the flesh as described in the previous section. Extending the procedure pursued in trying to get a clearer sense of the significance of the flesh, we now need to consider how these indirect features of orientation are given over by analyzing their silent significance within our everyday sense of the world. (This will be a procedure much like considering how the flesh manifests in the experience of riding a bicycle.)

To begin this process I have created a simple table. It describes orientation-toward and orientation-around in light of how these might be experienced from the subject side and object side of our experience. Of course, we must remain persistently aware that orientation belongs explicitly to neither our subjective nor objective sense but to the betweenness of the flesh that gives rise to this distinction. There is something inherently artificial about this way of divvying up our experience of orientation in the flesh, but it can, nonetheless, be quite helpful.

|  | *Orientation-Toward* | *Orientation-Around* |
| --- | --- | --- |
| Experience as Subject | Desire | Identity |
| Experience as Object, Space, and so on | Intention for work | Horizon |

In the initial description of the orientation-toward and the orientation-around, we have been implicitly working our direct experiences from the subject side. Orientation-toward manifests as a desire for some proximity. Orientation-around manifests as a sense of identity or being taken up by a concept that directs our future desires.

From the object side of our direct experiences, an orientation-toward is given through an intention for accomplishing some *work*.[36] The orientation-toward is given in terms of an identifiable action—an aim—by which the taking up of the object by a body (as an intertwining of flesh) can be considered successful or failed depending on *how* the object is taken up by the body of a given subject. In successfully performing the action characteristic of the orientation of the object, the body co-inhabits the object (the inhering of the flesh described previously). A fruitful orientation-toward would be one where the risk of exposure and advancement of the self as lover is able to realize the meaningful, dynamic possibilities—a flourishing—of the object through its intertwining. We see this in the example of riding my bicycle down the hill. The experience whereby my body is extended through the bicycle can be divided along these lines of orientation-toward. My desire for the proximate experience of riding the bike is realized and the possibility of the bike to travel down roads (its action or work) is accomplished.

From the object side of our direct experience, an orientation-around has the character of a horizon. No object appears in utter alienation; a host of other elements that form its horizon shapes it. The orientation-around characterizing the object is given by how the object shapes this horizon as it bundles together other objects—other objects that support the work of a specific orientation-toward—nearby. It is a tendency of things to appear together. With regard to my riding the bicycle, the orientation-around on the subject side of our direct experience is my desire to express some identity of "bike-rider" that encourages me to pursue my morning commute, intertwining me with the bicycle, even in foul weather. On the object side of the direct experience of orientation-around, I find that my bicycle riding is pursued with a host of accompanying accoutrements: a helmet for safety, sunglasses to combat glare from the road, gloves to help with my grip, and shoes that hook into my pedals. The accoutrements facilitate my intertwining with the bicycle, deepening our sense of inhering together and augmenting my claim to identity in the subject side of this direct experience of orientation-around. Additionally, it places a whole suite of objects in proximity to me—making them available or proximate as something I can be oriented-toward. Just as with the orientations-toward, these orientations-around are united in our actual experience of the flesh even if we might artificially break them apart by taking a perspective from outside this intertwining of flesh.

So far, we have dealt with instances where the chiasmic intertwining of the flesh occurs smoothly. These are instances where the dehinscence, or *écart*, shifts easily so that the intertwining of the flesh is robust enough to evoke a new way of understanding our being in the world and there is a harmony between the four facets of orientation described earlier. Not every experience of chiasmic encounter leads to this easy intertwining. One of the consequences of inheriting orientations and various proximities is that not all objects and spaces are equally or readily available to my body, and this bundling is hardly accidental or causal:[37] "Insofar as we inherit that which is near enough to be available at home, we also inherit orientations, that is, we inherit the nearness of certain objects more than others, which means we inherit ways of inhabiting and extending into space."[38] If this bundling is not accidental, then we must carefully consider the background conditions by which some objects, and not others, are being made available to particular bodies, because my extension into the world or incorporation of the world to my body relies on a certain proximity. Without it, the potential of having fruitful crossings of the flesh can become frustratingly stymied.

Phenomenology cannot restrict itself to descriptions of bodily incorporation that are easy; it must also describe *disorientations* and their potential significance. As Ahmed astutely notes, "A phenomenology of 'being stopped' might take us in a different direction than one that begins with motility, with a body that 'can do' by flowing into space."[39] How then do we characterize disorientation?

In Chapter 6, I indicated that disorientation was a slipping away of what seemed to be available. Understanding orientation and the flesh, we can now be more specific. Disorientation is a failing of the chiasmic formation of the flesh. What should have been an enabling intertwining, evoking a new and meaningful way of being in the world becomes a retreat;[40] the fleshly crossing fails to sufficiently actualize in some way. This occurs when there is a disjunction between the facets of orientation that are otherwise in harmony during a successful chiasmic crossing of the flesh.

We can begin to imagine all sorts of ways these conflicts of orientation might occur. Our desire may not align with the intention for work available through the various proximate objects with which we intertwine. Perhaps such orientation-toward is able to be chiasmically realized, but the new way of being opened by this orientation-toward is in conflict with the wider horizon of proximate objects or what constitutes a legitimate sense of identity. Perhaps we have a solid sense of identity, but our desires betray this identity we affirm, or there are no proximate objects in the horizon for us to seek to be intertwined with in the flesh so that we can make this identity our own.

The possibilities for how we might construct a typology wherein one aspect of orientation falls out of sync with the others are numerous. Establishing a formal litany of the ways these conflicts might arise is not my primary interest here. What is important is that in each case where some conflict arises in these orientations, there is a disruption of expectation. One aspect of our orientation in the flesh attempts to make proximate something other aspects of our being oriented in the flesh suggest to be unavailable. Most often, we might experience this as a desire (an orientation-toward some proximity) askew or beyond the existent horizon of proximate objects (the body of objects and spaces that can be oriented-around).

We cannot dwell forever in disorientation. Intuitively we feel this, and existentially we expect this. There is discontent and anxiety in disorientation; it is an obliqueness of the world that is disquieting—where the familiar becomes strange. The meaningful existence produced by the smooth, chiasmic encounters of flourishing flesh is disrupted as we experience the shock of falling out of attunement with the world; an anxiety where the familiarity of the everyday collapses around us begins to well up in conjunction with this disorientation.[41] We want to re-orient the flesh so that the attunement is felt again; we want to bring this dissonance into harmony and re-establish the sense of the fruitful intertwining between self and world that is flesh.

In presence to the flesh, disorientation registers—in its immediacy—as a wounding in our exposure and a shunning of our advance. If, as the *imago Dei*, we seek a meaningful ordering of existence that promotes new ways of creation's flourishing, the space we make of ourselves for some alterity to find its meaningful existence through us is experienced as unfulfilled or incomplete in disorientation. To remain so wounded and shunned is to have the very meaning of our existing crossings in the flesh thrown into question: Have we been oriented-around the flesh in a way that does not actually give us the assurance against vanity we have sought through crossed phenomena? Is disorientation simply an indication that we have irreparably failed in living-into shared responsibility and accountability for facilitating the flourishing of that which is alterior to us? Anxiety swells in considering that our orientation of the flesh has not yielded a sufficiently meaningful existence in light of this disorientation. We cannot remain in this state for long and continue to hold onto our particular enfolding of the flesh.

To get at this sense of disorientation (and how we might resolve it), we can think back to an astrobiological example: Kepler-452b. If we apply what we know of the flesh and orientation to the study of Kepler-452b, then this planet is never given over to us as a bare thing in-itself. It always comes to each of us as a particular entwinement in the flesh. This intertwining in the flesh between me and Kepler-452b is not instituted *ex nihilo*; this fold of the flesh exists in relation to a set of orientations that have been instituted through other instances of the flesh being intertwined with Kepler-452b and hosts of other exoplanets. After all, Kepler-452b is just one of *many* other exoplanets that could be considered as they are enfolded as flesh (one of 4,126 confirmed exoplanets as of February 25, 2020).

Returning to my own lived-experience of this exoplanet, I have to recognize my finitude; I cannot take up a limitless number of objects. Throughout the day, my lived-body is persistently incorporating the world around it—wrinkling the flesh. This leaves only so much time and space for the specific incorporation of exoplanets. The fact that I do incorporate Kepler-452b (in this very consideration of it), but not Kepler-296e or HD-142b, has something to do with how Kepler-452b is proximate to me in a way that these other exoplanets are not. Kepler-452b is oriented toward the flesh in a way these other exoplanets are not; it makes possible ideas or concepts these other exoplanets do not; it allows me to do a certain kind of work in the world when I am oriented-toward it that the other exoplanets do not.

None of the features of Kepler-452b's proximity are static. As is easy to imagine, Kepler-452b might not be incorporated into my lived-body beyond having a vague sense that "Hey, that was a famous exoplanet and I haven't heard about it in a while . . ." without the dedicated time and space that I am giving for constructing a richer orientation toward objects of astrobiological interest through writing a work such as this. As an interested non-specialist (a loosely designated way to characterize my orientation-around this astrobiological phenomenon), it is probably safe to say that I have a deeper desire to meaningfully encounter Kepler-452b (an ability to be oriented-toward this exoplanet) with a deeper sensibility for the possibilities opened by it (Kepler-452b's orientation-toward the flesh) as it forms part of a wider horizon of exoplanetary phenomena (Kepler-452b's orientation-around other planets) than a pure novice. However, my capacity to take up Kepler-452b and understand its orientations is certainly less than that of any specialists in this field. This should come as no surprise. Moreover, in an important sense (because the orientation is fundamentally a part of the body-object matrix), the phenomenon "Kepler-452b" itself is not the same across these various sorts of fleshly intertwinings.

It should be clear by now that the way in which Kepler-452b is "oriented-toward" in the midst of the flesh cannot be established by one person alone. This orientation-toward is always shaped by the vast history of repetitions that form the orientation-around Kepler-452b as it has been incorporated into the lived-bodies of others. I inherit the orientation of these incorporations (as for instance, I do not study much of the data about exoplanets directly but have it given to me through articles and particularly news media) along with the innumerable other aspects of orientation that accompany the other objects in the world that are proximate to me.

If it is the news coverage of Kepler-452b that has established its proximity for me (and much of the wider public), this proximity is thematized in terms of similarity. Thinking back to the first chapter, Kepler-452b was presented as your planetary cousin. It wants to borrow some plants for crops and give you a place to go tanning; sure, it may be unreachably far away, but in the ways that really matter (such as star type and location in the habitable zone), this exoplanet is supposed to be like Earth. The similarity allows me to project myself, or perhaps more specifically my everyday sense of taking up the flesh, into a new world. Kepler-452b is a confirmation that my place in the universe is not restricted to Earth. It has a horizon, an orientation-around itself, premised on similarity.

However, as more information is carefully assimilated or discovered (such as the significant decrease in probability that it is rocky, has an atmosphere like ours, or perhaps exists at all) the incorporation of Kepler-452b and its proximity to my lived-body is challenged if the horizon the exoplanet is oriented-around is constituted out of similarity. Kepler-452b falls into being an object of some disrepute; it gets dislocated from this horizon and now the orientation-toward—whether understood in terms of the subject as my desire or in terms of the object as what possibilities the exoplanet makes available—is stymied. Moreover, my identity (as a form of subjective orientation-around) is thrown into question (for example, "Have I been misunderstanding these exoplanets?" "Did I misread the news reports?" "Do I really know what *I thought I knew* about these exoplanets?" "Can I really call myself an 'interested non-specialist' or have I just been a novice all along?"). As the horizon disappears, Kepler-452b becomes disorienting; it disrupts the other aspects of orientation reversibly related to this one and they are thrown askew.

In my anxiety over this sensed disorientation, I might "let go" of Kepler-452b. I cease to pursue any sort of fleshly intertwining with this exoplanet and do not incorporate it into the space of my lived-body. The orientation-around similarity is sufficient and this particular exoplanet is no longer proximate to me as I once thought it might have been.

Yet this solution should rub us the wrong way. It cuts against the presence of the flesh that I advocated for in the previous section as something to foster if we are to become the *imago Dei* together: a continuous exposure and advance that makes me available to this exoplanet such that it finds a way to meaningfully flourish in the flesh. Such a retreat from intertwining is dissatisfying. The flesh, in all its complex relation and reversibility, seeks new ways of being knotted up in the world. To enrich our chiasmic encounters—especially ones that seemed impossible in our prevailing orientations—presents new possibilities for entwining the flesh; to enrich these encounters is to enrich the field in which our lived-bodies extend, in which we are exposed, toward which we advance, and through which we realize the depth of the meaning of our being. There is something disappointing about retreating from such a possibility. More starkly, if our accounts of planetarity and the *imago Dei* hold, then such a retreat from intertwining may shirk our right to be responsible and live into a human way of being. To "let go" of Kepler-452b is to give up on refracting the creative power of God such that this exoplanet has a meaningful existence in the flesh.

To call this letting go of Kepler-452b a shirking of our responsibility and potential to live into being the *imago Dei* may seem a little farcical: It is one exoplanet very far from here and, as I have already noted, as finite creatures there are only so many spaces and objects that can be maintained in proximity. Shouldn't we just let this one slip off the radar?

Instead of letting go of Kepler-452b, we might assuage our anxiety in the face of this experience of disorientation by reframing the horizon of similarity that conditions the way we are oriented-toward this exoplanet. We maintain our desire for enlacement with Kepler-452b; we affirm its proximity even as it seems to slip out of the horizon of similarity. This disorientation is uncomfortable, but if we dwell in it—sustaining our crossing with Kepler-452b in its obliqueness to our expectation for meaningful contact with exoplanets characterized by similarity—we can begin the hard work of building a new sense of being oriented-around exoplanets. This means constructing or shaping a new horizon not premised on similarity.

This example may be innocuous in its outcome for most of us. Yet, how we resolve the anxiety experienced in facing disorientation matters deeply. The attitudes and habits we adopt in facing innocuous instances of disorientation will shape the way we can confront more serious disorientations—especially those posed by the symbol of the Anthropocene.

To reframe the horizon and maintain a crossing in the flesh with Kepler-452b instead of cutting off this crossing is to adopt a queer orientation or to make Kepler-452b into a queer exoplanet. It is queer in that it is oblique to our inherited orientations about what constitutes a meaningful exoplanet. More generally, a queer orientation is one that wanders from the typical or "normal" course shaping the way in which we can orient-toward or desire something: This queer orientation is askew or slantwise to a prevailing normative orientation-around—a perturbation of the direction of the normative orientation.[42]

What makes this difference? How do we account for choosing to assuage our anxiety in the face of disorientation through a straightening versus remaining with—lingering and loitering—in the disorientation so that we do not abandon a crossing of the flesh but instead queer it? What mood must we adopt toward queer "disorientation devices" such that we persist with the anxiety of disorientation and allow this obliqueness "to open up another angle on the world" as a full crossing in the flesh instead of an abandoned possibility left fallow?[43] What attitude must we affect so that we develop a disposition toward presence in the flesh that might pause even at the dissonance of disorientation in order to search out hidden harmony opening us to impossible possibilities for being-in-the-world?

For this, we need wonder.[44] Wonder is a mood or attitude that we take in the face of realizing that our everyday assumptions are untenable. It is a sort of awareness and surprise at what we do not know—the very abstruse quality of their being an everyday world—that forms the ground from which our thinking as such springs. It is the feeling of being aghast by our very existence. As Mary Rubenstein puts it in analyzing Plato's *Theataetus*, "Wonder, then, comes on the scene neither as a tranquilizing force nor as a kind of will-toward-epistemological domination, but rather as a profoundly unsettling pathos . . . [T]he philosopher's wonder marks his inability to ground himself in the ordinary as he reaches toward the extraordinary."[45] Wonder is the attitude we take in opening a rift in the midst of the everyday world, a rift that, when closed, marks the cessation of wonder.[46]

Wonder (*Erstaunen*) is *not* mere curiosity of any kind. Martin Heidegger provides a helpful catalog of curiosities in this regard so that we first make sure to understand what authentic wonder cannot be. It is not a curiosity exhibited toward the comprehension of new novelties (*Verwunderung*). Nor is it a fervent curiosity that persists beyond the initial experience of novelty to seek out a deepened comprehension of some set of unusual complexities in the everyday world (*Bewunderung*). Nor, finally, can wonder be a marveling at one extraordinary thing to the exclusion of our experiences in ordinary everydayness (*Staunen* and *Bestaunen*).[47] All these are modes of curiosity that seek the surety of knowing, not the unsettling pathos of wonder that stands open before the enigmatic quality of our very existence. In wonder, we give attention to the mysteriousness of everydayness.[48]

In light of our consideration of disorientation, we can think about the distinction between curiosity and wonder as two different means to dealing with the unsettling anxiety of disorientation. Curiosity alleviates our anxiety in constructing knowledge that allows us to comprehend the disorienting phenomena and situate it amidst existing categories of our understanding. Wonder remains with the disorientation and the anxiety of our conflicted orientations that come with such a chiasmic crossing in the flesh. It remains in the inscrutable crossing of the flesh and holds onto this experience of anxiety.

We remain because in wonder we await a double movement from this experience of anxiety. There is terror and awe. There is terror in the face of having our sense of everydayness upturned by this rough crossing of the flesh. The orientations of the flesh that we have previously so easily adopted become troubled; the chiasmic crossings of the flesh once so certain now cannot possibly be. But we can remain with the terror because we are full of awe at how this disorientation provides a new vantage on the world; when we stay with this disorienting crossing in the flesh "a space is cleared for a new arrival."[49] This crossing in the flesh happens where it ought not have and puts us back into the midst of the world in a new way. The old possibilities are torn asunder as the new impossible possibilities—new ways of flourishing and affecting the flesh in meaningful existence—are evoked.

In a wondrous attitude, the mood whereby we take up the flesh in wonder, we expect this double movement. Thus, even when we encounter crossings of the flesh that are disorienting, this is not something to be lamented. Certainly, we do not seek out the anxiety this disorientation produces, but when we experience disorientation as we seek out new chiasmic encounters of the flesh, we take solace and have courage in the hope that the double movement of wonder promises. Consistently adopting this attitude, we can develop a disposition of presence to the flesh. We can bear the risks of self-exposure and the advance that makes our bodies into a space for the flourishing of the flesh when we have wonder's surety: The queer, disoriented taking up of the flesh that provokes the most anxiety will still lead us to some hitherto unforeseen meaningful way of existing and flourishing as flesh. The terror of anxiety is only ever half the story if we wonder before the world.

## Play

If persistently adopting an attitude or mood of wonder can lead us to a disposition where we are present to the flesh, then we must now reflect on how we foster this wondrous attitude. This is no small question. Why do some readily adopt the courage and solace of wonder's double movement in the face of anxiety while others make recourse to controlling categories of curiosity? What *praxes* might lead us to be more likely to experience the world in terms of a wondrous attitude?

Whatever *praxes* might develop our sense of wonder, the mutability of the flesh will be very important in them. The examples I gave in thinking about wonder were of more long-lasting and fundamental change to our everyday assumptions: queerings of the flesh with deep and pervasive effects. However, the running example I gave of flesh in the first place was riding a bicycle, where I emphasized its mutability and dynamism. To deal with the more long-lasting changes of queering flesh through the double-movement of wonder, we need to build up a surety in the hope of wonder's second movement—awe at new ways of flourishing in the flesh made available. We cannot simply be thrust into the most extreme cases of terror inducing fundamental change to our way of being in the world and expect that we can bear this anxiety in wonder. We have to feel out this double movement in settings with lower stakes, with crossings of the flesh that can be taken up and put down so that we can explore how persisting through disorientation might bring us to the awe of new ways of flourishing.

Playing can do this particularly well if it is freed from the everyday, binary sensibility that most often comes to mind. In its binary sense, we tend to define play as a "not": not-serious, not-work, or not-earnest. It is frivolity or leisure in contrast to the serious affairs of the everyday world. Yet, as Johan Huzinga aptly describes, play "is of a higher order than is seriousness. For seriousness seeks to exclude play, whereas play can very well include seriousness."[50]

To see this, however, we have to get at play *indirectly*. Play is a mode of being that does not directly reveal itself in the subjectivity of a player. One has to find how play appears through the intra-action of the players because it is persistently in-between. In terms of the language of flesh and presence above, we can say that the flesh is made present in play. The exposure and advance that open us to understanding how we are intertwined with others in a primordial way (*Ineinander*), giving otherwise unassailable meaning to our existence, is itself a sort of playing together (*Ineinanderspiel*).[51]

We have to put together the traces of play from instances of playing. More specifically, "conveying the meanings of playing also involves evoking memories of playing, which requires description and characterization. Only then do the words (as symbolizations of playing) join with the embodied memories of playing."[52] To understand play, we have to get into the depth of this embodied memory and so it does not lend itself well toward a direct characterization.

Because of a certain implicit awareness of playing's resistance to direct characterization—being a phenomenon not well possessed by the player but happening between our usually well-defined categories—what we tend to find in literature, particularly on animal play but also in child development literature on play, are catalogs of descriptive dimensions of playing. The elements of these descriptions, when taken in various combinations, should identify when play occurs. How these catalogs work to form a definition is not always obvious or consistent, though. Here, the aim is slightly different. I want to pick out certain family resemblances across the descriptive catalogs of what constitutes playing such that we gain a better sense of what playing *does* and why that might matter for developing a wondrous attitude before the world.

To that end, we might think about how the descriptive catalogs of play implicitly are trying to provide us with answers to the who, what, when, where, why, and how questions learned in elementary school as a way to start investigating and reporting on an event. Who usually does this playing? What exactly do I do when I play? Where and when does play happen? And, how does playing matter—why play in the first place?

We might start with the where and when question because it gets at a vexing issue of play: How does play relate to the "real world"? Alfred Schutz suggests the world of daily life is the world of our everyday anxieties. It is the place we enact the practical and pragmatic interests of our survival, responding with startling precision to a deductive logic: If this, then that. The chain of this sort of thinking almost seems stilted in its simplicity, but it is very effective. If I do not eat, then I will die. If I do not gather food or farm, then I cannot eat. I do not know how to gather food or farm well, but I still need to eat. If I barter something else I have of value, then I could get food from someone who does farm or gather. The basic idea here is easy to follow. At some point, perhaps because of its ability to serve these practical interests or the reliability of this deductive structure, this world of daily life gets taken to be reality.[53]

Play, however, does not directly contribute to our survival in any way that fits simply into this deductive structuring; in this sense, where play happens is always at some distance from reality. However, this distance also indicates a certain dependence. If I am dead, I cannot play. Play may not directly contribute to the survival envisioned by this "real world," but it cannot happen without meeting the basic requirements of the "real world." Animal play studies recognize this requirement and indicate that play behavior has to be initiated in the context of an evolutionarily relaxed field. In short, we can play when we are adequately fed and free from various competitive and predatory stresses.[54] In these relaxed conditions, one is made *free* to play.

An evolutionarily relaxed field may provide the conditions of freedom for play, but the freedom itself implies something much wider and more complex about how play happens. Playing is not something that can be forced or mandated—it cannot be required.[55] When I was bored or frustrated with some activity as a child that was meant to be fun, my father would sternly look at me and say, "We will play and you will enjoy it!" His faux command (almost) always made me laugh a little, lightened my mood, and ironically opened me to a sense of freedom that allowed me to start playing again. I find myself intoning the same farcical command to my own son now with a similar effect.

We might be drawn into a game or enticed into playing, but ultimately it is only in freedom that we can step out from the real world into the world of play. We make this choice in beginning to play. It may not be a conscious choice (we may simply slip into playing), but the choice to leave the real world for the world of play cannot be mandated or forced: I cannot impel you into adopting a state of play. To quickly summarize, we might say that play happens apart from the real world (where) but only after the conditions that satisfy the real world have been met (when), thereby relaxing the field of possible actions I may take in the world giving a freedom that can be actualized as play.

We must take immediate caution not to overemphasize playing's distance from reality, though. When play's distancing from reality becomes escapist, play takes on the quality of an illusion and loses critical facets of the wider importance and conceptual power it might have.[56] If we emphasize that play is distanced from the real world or daily life, we must do so always remembering the *significant* caveat that daily life is not the only reality. As Robert Bellah realizes, "In spite of its 'apparent actuality,' the world of daily life is a culturally, symbolically constructed world, not the world as it actually is."[57] If we bear in mind this constructed quality of the real world, then play (along with states like dreaming, leisure, art, science, and religion), according to Bellah, helpfully enables us to imagine new and different realities that contrast the apparently "real world" of our everydayness. In these alternative worlds, everyday assumptions no longer rule.[58] It is in this sense that play is at a distance from the real world. When we are in a state of sufficient freedom, play can occur as a way of imagining alternative realities, new worlds, and assumptions contrary to what we have received. This is certainly *not* to claim that play lacks importance for the real world of survival (this is a staple feature of studies of animal play); it just does not take place in the "real world" of survival.[59]

Ascertaining how to describe play's potential effects on the real world is a point we will have to return to in a moment. For now, I want to shift toward the "what" questions. What happens when the conditions for play arise? What is it that we recognize as playfully occurring? What goes on when play creates these alternative realities? At the very least, we might begin by noting that play has an imaginative quality. In play we engage a transitional space of possibility—an "as if" space. We leave the literal quality of the real world, suspending aspects of its rules and expectations, for this alternative version that play makes possible.[60]

Technically put, this imaginative capacity is a counterfactual engagement—a possibilizing. Alison Gopnik makes the idea very accessible: "Counterfactuals are the woulda-coulda-shoudas of life, all the things that might happen in the future, but haven't yet, or that could have happened in the past, but didn't quite. Human beings care deeply about those possible worlds—as deeply as they care about the real actual world."[61] This ability to imagine counterfactuals lets us consider alternative ways that the world could be.

Counterfactuals may be a technical way to describe this, but really playing is a way of pretending or imagining. Pretending is not meant to be a denigrating term; pretending allows us do many things. By pretending—invoking these counterfactuals—we can plan, decide between alternative possibilities, and reflect on previous actions so that we might change how similar events proceed in the future. Certainly, these are some evolutionarily advantageous skills.

It is important to bear in mind that this pretending is not pure fantasizing. Pretending and imagining make much use out of aspects of our ordinary lives, but in the play world, those aspects get displaced from their function and location in the real world. The dislocation allows us to create new associations and uncover facets of these adopted behaviors that might otherwise remain hidden; we construct alternative causal maps through playing.[62]

Here we might make a connection back to the orientations of the previous section. If in playing we pretend and imagine, then we invoke counterfactuals that make new relations possible. In play, we intentionally take on orientations-toward aspects of our world that are unexpected or discordant with the more widely available horizons of our current way of taking up the flesh. This is precisely a *disorientation:* one aspect of the orientation of the flesh falling out of line to the others.

The imagining and pretending of play produces disorientation. However, because it is removed from the seriousness of the real world, this disorientation is less threatening and we can more easily tolerate not straightening its queer effects right away. Quite the opposite, we encourage dwelling in this sort of disorientation and call it fun. In fact, without dwelling in this disorientation, something about the imagining and pretending becomes less than playful: telling us something important about how play happens.

Play has an immersive quality. I have already emphasized that the alternative worlds of play are not escapist fantasies. Yet in play there is a performative or representative dimension that Huizinga calls a "stepping out."[63] When playing, we take on a role different from the everyday. Children do this both alone and with each other all the time. My son becomes Ash when his friend decides to be Pikachu so they can battle an imaginary Team Rocket; my daughter has repeatedly saved Montanui as Moana; and our laundry baskets often become cars so our children can drive around town together during laundry day.

This immersive sense of being seized by our role in play is not limited to these direct acts of pretending in young children. Our teenage neighbor becomes the second baseman for his high school baseball team, vigilantly looking for the next ground ball hit in the game, as he practices outside and calls his own color commentary. Even adults continue this immersive play; personally, I still relish having time to step out into my area of hexagons in Settlers of Catan.

In each case, to engage in immersive play is an act of *poiesis.* It is a making real of some new phenomenon for all the players participating. The immersive play becomes a performance that brings a way of being in the world and a particular world of being to life.[64] We cannot opt into or out of the creative process of play. Play relies on our immersion into the new limits, rules, and boundaries (whether spatial, temporal, or causal) that make up the imagined counterfactual space. It is only by abiding by these rules of our new world that it can come to be. Play is its own sort of sustained, immersive effort.

If there is a difference to highlight in the examples of immersion in the preceding text, I would suggest it has less to do with the intensity of the immersion (each case requires the players to commit themselves to the world of the game) but with the threshold held for tolerating the counterfactual quality of the particular reality. How far will we let our imaginations run—how many rules of the real world can we suspend—before this wild immersion in the reality of the pretended collapses on itself? Children's pretending can almost seem unhinged to an adult. Gopnik suggests this also, simultaneously indicating why it can seem like children play more frequently than adults do:

> From the adult perspective, the fictional worlds are a luxury. It's the future predictions that are the real deal, the stern and earnest stuff of adult life. For young children, however the imaginary worlds seem just as important and appealing as the real ones. It's not, as scientists used to think, that children can't tell the difference between the real world and the imaginary world. . . . It's just that they don't see any particular reason for preferring to live in the real one. . . .
>
> Our young selves get to freely explore both this world, and all the possible counterfactual worlds, without worrying about which of those worlds will actually turn out to be inhabitable. We adults are the ones who have to figure out whether we want to move into one of those possible worlds, and how to drag all our furniture in there too.[65]

Gopnik makes clear that adults encounter fictional worlds as a luxury. There is a concern for meeting the needs for survival in the "real world" that makes these fictive worlds lagniappe—a little something extra permitted if we meet our needs in the real world. Children traditionally do not have as many—or sometimes any—responsibilities for meeting the needs of their survival in the wider environment and this frees them to set their real and imaginary worlds on equal footing.

Additionally, Gopnik implies that because the precedence and concern associated with the "real world" and its deductive logics of survival are suspended for children, they can tolerate wider counterfactuals. The rules of what constitute a habitable world are already looser for children than for adults. As a result, it is not surprising that they exhibit a freer tendency than adults to create immersive worlds that remain believable to them but might be simply fantastical for an adult.

In any case, it is important to remember that the creation of these immersive, alternative worlds is not permanent, no matter the age of the players. If the worlds were permanent, they might rival the everyday world itself in constituting the real. Certainly, one goal of play might be to return from the world of play and change the everyday (as I would wholly endorse), but that does not mean that play is permanent. Moreover, the impermanence is hardly a problem for at least two reasons. First, impermanence is itself part of what gives play its potential power as a disorientation device (because we cannot dwell forever in disorientation without crippling anxiety). What matters is that the impermanence does not disrupt the flow of play and the sustained effort made to abide by the limits, rules, and boundaries of the alternative play world. The impermanence of play only becomes problematic when we do not sufficiently immerse ourselves in the alternative reality of the play world.

Moreover, the impermanence allows for repetition. In repetition, play reveals itself to be a "use it or lose it" skill. When you do not play, playing gets harder. This point seems banal, but the elements of habit and repetition implied by play have long been a critical philosophical tool and also serve as a critical theme in the study of animal play. Burghardt observes that play involves imperfect repetition.[66] In repeated bouts of play, the possibilities for the play behavior are explored and honed. Play like other skills requires repetition for its enhancement. We might even say that with practice we get better at play.

We can imagine our one-line examples from the preceding text in this regard. My son's pretending to be Ash gets more involved each time. Even over the course of a few days playing Pokémon outside, the game went from having all of his friends pretending to be the lovable Pikachu to including a richer cast of characters—Charizard quickly became more popular. Over the course of a month, characters I did not know about from our reading together were making appearances, and he and his friends were imagining new Pokémon altogether. The process was not simply additive, though; they excluded some characters from their game as well (often because the logistics of being a character violated an implicit rule of this counterfactual reality). In all this, the alternative reality and its rules became more complex; his immersion in this alternative world of play became increasingly thorough as the repetition made the play reality increasingly rich.

The same process of repetition breeding further immersion, skill, and complex understandings of the play world can be constructed for the other examples. Recently, I saw my daughter laying out newspaper in front of her laundry-basket car to be the road she was driving on, and then she started shouting at an imaginary driver next to her to stop talking on his cellphone! The role of practice is easy to imagine in the case of our neighbor, the second baseman. Physical skills are developed through practice and playing in games. He becomes more adept at fielding and batting, but with practice, he also comes to understand strategy. Now the game can involve safety squeezes and cutoff men. Even in my own love of board games, I develop such skills and strategies. I become better at interpreting what the other players might do in a given situation and changing my strategy as the chance aspects of the game unfold. I become more immersed in the subtler nuances of the game and the real world falls away further than during the first time I played.

The repetition enhances the *poietic* power of play both at the level of the skills developed (the seemingly all-important future orientations of adults that Gopnik identifies in the preceding quote) and in terms of the depth of the alternate realities imagined. It is no wonder that children might prefer living in these alternative realms to the everyday world predicated on usefulness. If we have the freedom and a relaxed field to engage in play, then these alternative realities create alluring possibilities.

In light of these features of playing, someone who refuses to enter into the space of play becomes a threat to the creative power it can hold over all the other players as well. Huizinga describes this with his image of the "spoil-sport":

> The player who trespasses against the rules or ignores them is a "spoil-sport." The spoil-sport is not the same as the false player, the cheat; for the latter pretends to be playing the game and, on the face of it, still acknowledges the magic circle. It is curious to note how much more lenient society is to the cheat than to the spoil-sport. This is because the spoil-sport shatters the play-world itself. By withdrawing from the game he reveals the relativity and fragility of the play-world in which he had temporarily shut himself with others. He robs play of its *illusion*—a pregnant word which means literally "in-play" (from *inlusio, illudere* or *inludere*). Therefore he must be cast out, for he threatens the existence of the play-community.[67]

While casting out the spoil-sport may seem extreme—it brings a certain *Lord of the Flies* vision to mind—the point is important when we start thinking about why it is that play matters. In a rudimentary way, there is an easy answer to this question. The skills practiced through play can develop into skills with evolutionary importance. The immersive practice that is characteristic of play sharpens skills that might then be more directly useful in our everyday struggle for survival. This may be the more familiar way we think about the significance of play and it is not unimportant. It emphasizes the role of repetition in play and requires immersion of all the players in the game so that the skills practiced are developed in earnest. The skills practiced need not merely be physical (for example, hunting, sexual display, and so on); they may also be social (for example, competitiveness, cooperation, and so on) and mental (for example, strategy, planning, and so on).

Play also has an ontological significance, though—particularly for an elemental or poietic ontology like that of the flesh.[68] In playing, we create alternative, immersive realities that belong between the players, spaces, and objects making up the field of the game. Certainly, play is temporary and it explicitly leaves behind an overwhelming concern for survival, but it is not overstating the case to indicate that in play we engage in a praxis of dwelling in alternative ontological frames.[69]

The flesh provides a ready ontological structure for describing this. Imagine that acts of playing are specific crossings of the flesh that in the time of their impermanence constitute our sense of reality. In the example of the bicycle, we get this sense; it is an impermanent but real sensation of my lived-body extending through the bike. We might call the crossings of the flesh with their orientations (yielding to the various incorporations and extensions of the lived-body into objects and spaces)—this poietic, elemental ontology—an ongoing act of play. Play relates immediately to developing presence in the flesh insofar as play is an intentional way of enacting crossings of the flesh that, in everydayness, seem to be at the very peripheries of what is possible. Playing helps remind us that while a deductive logic of survival may constitute the everydayness of the "real world," this is not the only reality.[70]

What is critical, as Hans-Georg Gadamer recognized, is that these other playful realities—realities illustrating the dynamic potential of the flesh—*matter*. The ontological world opened by play points to a truth simply unavailable in the "reality" of our everydayness; as such, the truths play opens before us can then transform our sense of everydayness.[71] In particular, play opens onto an experience of the arising of self and world unconditioned by the serial quality of the 'real world.' We are not simultaneous when playing: co-existing in our individuality with the things of the world passively available for our use.[72] Instead, play gives us the experience of being formed in a meaningful whole—an immersive alternative way of being, that transform the serial sensibilities of the real world into a persistent contemporaneity. This allows our everydayness to be transformed from the self-experience of individuality to the chiasmic touch of being-with where our sense of bodies, objects, and space co-institutively emerge by their intertwining. With repetition, this intertwining can provide a sense of meaningful existence beyond the persistent sense of stark individualization as we are weaved into the crossing weft and warp of the flesh that we indirectly sense.[73]

The repeated pattern of this experience of gaining a sense of self in the playful realities of alternative fleshly crossings—playing again and again—prepares me to adopt an attitude of wonder. In play, the loss of the everyday world is not terrifying because I am already suspending the reality of the everyday by playing. Play intentionally brackets the first movement of terror and anxiety in wonder. Thus, I am spirited along to the second movement of awe, finding an otherwise impossible set of possibilities meaningfully made available to me and explored in the disorientation of play. Play develops a proclivity in me to expect the possibility of new ways of meaningful existence to emerge from these disorienting, fleshly transformations.

### Becoming an Artful Planet

The first seven chapters make a distinct case for interpreting crucial symbols related to human being, the *imago Dei* and the Anthropocene, with astrobiology's intra-action of living-systems and habitable environments forming the context of engagement for such interpretation. I developed an account that described meaningful human being as a particular kind of intra-active living-system: a refractive system opening up the creative power of the divine into new forms of meaningful existence and flourishing within creation. As such, we are an inextricable part of Earth's planetary systems, shifting biogeochemical cycles to technobiogeochemical cycles, affecting a distinctive agency that shapes the very conditions of planetary habitability. *Human being is the living-system that brings about*

*technobiogeochemical cycles in its intra-action with the habit-*
*able environment such that new ways of flourishing and meaning-*
*ful forms of existence are made available to other living-systems.*
The *imago Dei* facilitates the transition by which a living planet
becomes an artful planet.

At the start of this final chapter, though, I also suggested that
astrobiology pushes us toward an ontological awareness: a need for
ontological language that can deal with the betweenness of astrobiological phenomena (i.e., the intra-action of living-systems and the
habitable environment) and the scale of its biological concerns (i.e.,
a concern for living-systems more than individual organisms). To
do theological or philosophical work in light of these inescapable
themes, we need a suitable ontological language to deal with intra-
action and the felt tension between our experience of individualization as a specific living thing and our meaningful being as manifest
through participation in a wider living-system.

This need for ontological language arises because we do not have
direct experiences of being an intra-active, living-system; we do not
experience this artful planet of the *imago Dei.* Our direct experience
is perspectival—a first-person distinctiveness that at its best might
understand itself as participating in the midst of this vision of the
*imago Dei.* We participate in a living-system and we participate in
instances of intra-action, but these are not possessions of a distinct,
individuated self in the way we tend to account for our experience
in everydayness.

If we interpret the meaning of human being as conjunctive with
living into this vision of an Anthropocene *imago Dei*, then we (perhaps oddly) do not have any direct experience of what it means to
be human in our everydayness because these concepts of deepest
meaning are not directly applicable in the existential experience
of individual human beings. We need an ontological language that
frames our lived-experience in a way that does not negate the individuality of our existential experience but opens that individuality
to an animating depth of intra-action. Without this, we risk a dispir-
iting heteronomy where participation in meaningful human being
is only affected by a negation of the individuality of our existential
experience—a sense that we are some unimportant cog in the wider
machination of the human living-system's development of technobiogeochemical cycles.

The flesh can provide the needed ontological language. As a form
of elemental and indirect ontology, flesh persistently points us

toward the primacy of intra-action stressed by astrobiology. There are two senses of the indirectness of this element, in particular, that I want to emphasize: its mediateness and its communication. The flesh serves as an emblem for a way of being in the world; it is the trace of Being that is revealed by a careful consideration of the gaps and intertwinings that occur as a result of our fundamental reversibility. The flesh is the formative medium of these intertwinings (a transcendental condition from which the everyday experience of subjects and objects emerge) that is never directly experienced, though it comes to light through these: the mediate experience of Being through the intertwining of beings.

Since the flesh is not directly experienced (it manifests as a mediate element of the intra-action between beings), the experience of subject and object, self and world, living-system and habitable environment in their polar tension never disappears. I will always work toward the flesh from the position of an experience where I am a self, held in distinction to that which is not myself. However, the whole polarity—the intertwining without the collapse of distinction—provides a richer ontological point of reflection than the simpler component parts of self and world. The flesh forms a more primordial ontological unit even if it can be divided into what are clearly identifiable and simpler parts.[74] Its realization is always conditioned by our intra-actions; the flesh is rendered as a trace to us through the myriad incorporations and adventings of self and world. It is like a connective tissue, a nonseparable difference, that cannot be ruptured (even if we do not perceive it to be there) but is constantly being reshaped and formed as part of the relational dynamic. Though we might expect this position in the flesh to alienate our own sense of meaningful existence, it does not. To be in the flesh is never to have the distinctiveness of my "self" erased; I am not made unimportant or simply replaceable like a cog in the machine of the flesh. My positioning in the flesh is distinctly my own and the flesh itself is only manifest through these direct experiences of a self and world, even if it makes each term highly dynamic. Yet the sense of the flesh being "my own" is never autonomous.

Because it is not static and always given over in this mediated fashion, the flesh has to be communicated indirectly; it is realized through a personal appropriation that direct communication is incapable of conveying.[75] We cannot simply tell someone of the flesh and assume he or she will immediately take it up; in fact, given the counter-intuitive quality of the flesh in light of our everyday sensibilities; we might even be suspect of a person who so readily adopts the flesh after a bout of direct communication. Instead, this ontological supposition has to be communicated as a felt possibility. It has to be made available to other persons through their own realization of a personal participation in the dynamic and poietic processes of participation in this element—of taking up such a way of being-with-in-the-world.

How then does the flesh—as an ontological language suitable to the intra-action of the astrobiological context of engagement—add something to the way we interpret the symbols *"imago Dei"* and "Anthropocene"? It helps us narrate how we might personally participate in these broader senses of meaningful human existence that does not lose sight of our individuality but constantly forces us to question any sense we might have of being stably autonomous.

Yet we cannot have these experiences without a certain *receptivity* to the indirect communication of the mediateness that the flesh represents. We have to have a felt sense of empathy that allows us to become entwined with that which is alterior to us. We need to generate a sense of personal participation by enriching our receptivity to shared feeling and shared understanding, a sense of *Einfuhlung* and *Eindeutung*, which makes our participation in human being as an intra-active living-system into more than a mere intellectual assent. We need a greater openness to feeling the "we is" (as with my example of the incorporation of the bicycle) that is elucidated by crossings in the flesh. Merely knowing of the flesh, uncovering its silent operations, is not enough. The three concepts focused on in this final chapter—presence, wonder, and play—speak to this issue of receptivity that remained undeveloped in the analysis of the first three chapters. They are what individuals do and should foster in one another such that we could more deeply live-into a planetary *imago Dei* as a technobiogeochemical agency fostering the flourishing of creating in its wholeness.

As a first step toward generating this receptivity, I suggested the theme of presence. Presence names the disposition or perpetual orientation-toward the flesh: an attunement to the possibilities of Being in the flesh that arises in light of my particular incorporating and adventing of the flesh. This disposition is one of exposure and advance that affects a being-toward the other. In presence we declare, "Here I am." We riskily open ourselves to otherness that gives the meaning of my being only in tandem with making myself into a space that supports figuring the flourishing and meaningful existence of the other. Through engaging this process of presence, my being is given over to me as it is *meaningfully alterior*. Another cannot be if I am not, just as I cannot be if another is not. In the presence of the flesh we understand ourselves to be dependent on others and others to be dependent on us. That mutual dependency gives an assurance that bare accounts of being, as atomistic or banal materiality, cannot offer.

This indirect communication of the flesh, that it cannot simply be made known to another but has to be taken up in presence in order to be understood, presents a vexing problem: How do you communicate this ontological element such that another takes it up in their own lived-experience? We might be able to replicate the idea of the flesh to them (and presence dictates that we should always treat the other—however diversely figured—as if they, too, will approach the world in presence), but there is no direct way for me to reproduce my own lived-experience of the flesh in their lived-experience. It has to be taken up as their own distinctive place of enlacement with the world in light of their empathic receptivity in presence. How, then, do you foster a disposition like presence?

To address this issue, we turned to thinking about the orientations (more classically in phenomenology what we might call intentionality or aboutness) that accompany the incorporation and adventing of the flesh. Drawing on queer phenomenology, we distinguished an orientation-toward and an orientation-around that could be applied to the ways flesh gives rise to a sense of self and world. These orientations are contemporaneous in the flesh, but in our direct experiences of subject and object they can be parceled apart. Roughly put, the orientation-toward of the subject was understood as desire; the orientation-toward of the object was understood as an intentionality for work; the orientation-around of the subject was understood as an identity or tendency for the flesh to be taken up in a certain way; and the orientation-around of the object was understood as a horizon or proximity in which things are to be found together.

When we smoothly take up aspects of the world—when the flesh is enfolded without friction thus unfolding new meaningful patterns of Being—there is a harmony exhibited between these orientations. Our desires correspond with the availability of objects in the world—they extend our productive and meaningful possibilities giving identity through repetition and gathering into a horizon to further smooth the continued means of taking up the flesh in this way. In these moments, the flesh is a silent depth that could easily be passed over. It has been there all along, but the connection is so seamless it is as though it becomes invisible.

It is in the moments of disorientation, an incoherence arising in the midst of these four aspects, when the flesh cannot simply fade into invisibility. What should be incorporated and advented through this encounter in the flesh affects a retreat and the flesh becomes strained—its intertwining stymied. The expanding intra-action of the flesh fails: My body cannot take up some aspect of its world that seemed so proximate and this particular tendency to incorporate the world becomes blocked; or, we might express this in reference to that aspect of the world that cannot accomplish the new work it makes possible and falls out of its place in a wider horizon. There are real stakes in these moments of disorientation because they indicate a blocking of the flesh; in disorientation some either previously or seemingly proximate way of being-with-the-world has come into question.

Disorientation is not a state we can forever remain in; we long for the harmonies of well attuned orientations in the flesh. Disorientation generates a state of anxiety as our desires and ways of imagining objects working in the world are in conflict with our sense of identity and the wider horizons of the world. There is disjunction between my orientation-toward and my orientation-around. The anxiety is felt as fear over my misjudgment about what are legitimate proximities in the flesh with which there can be crossing or the sense of everydayness shaped by our orientations-around being overturned.

In our rush to re-orientation, we might discard such an offending crossing of the flesh. We let go of the troubling intertwining suggested by a particular orientation-toward and turn away from the feeling of terror that arises at the prospect that my take on the world has been poorly attuned. Alternatively, we might linger in the anxiety over disorientation, bringing us to a sense of awe that this crossing in the flesh might open a new angle on the world wherein means of intertwining the flesh are made possible that were otherwise impossible. The world becomes proximate to us in unheralded ways.

In wonder, we expect this double-movement of terror and awe in the face of disorientation. We have hope for awe as a continuous possibility on the other side of anxious terror. Without this wondrous attitude our presence in the flesh would always be only partial: only extending to those intertwinings of the flesh instituting harmonious orientations. Wonder opens presence wider; in wonder the availability of presence to the flesh, whereby we make ourselves a space for figuring the flourishing and meaningful existence of the flesh, is extended to dissonant orientations that are otherwise cast off if we do not have sufficient courage to face the terror of wonder's first-movement. But, wonder requires developing a propensity for lingering in disorientation: an expectation for the awe-full hope of impossible flourishings of the flesh. We need a capacity for persistence and resilience in the face of disorientation to adopt a mood or attitude of wonder at the world.

These are capacities developed in play. The alternative world of counterfactuals that we create in play are imaginative refigurings of how the flesh might be shaped in its various crossings. We might say that to play is to intentionally become disoriented; but, the disorientations of play are always low-risk because they are temporary and imagined through effecting a specific distance from the "real world" of survival. Play only happens in freedom and relaxed conditions that allow us to explore possibilities that take us well beyond the everyday world understood in terms of the work of survival. In a sense, play is the safe-space for disorienting the world.

So, in these immersive but bounded periods of play we create alternative worlds that would be disorienting in our everydayness. This allows us to condition an expectation for awe while circumventing the full feeling of terror at the prospect of overturning our orientations-around the flesh that shape the everydayness of our world. To make such a claim, we must recognize though that the distance from the real world affected by play is not some simple escape from everyday reality; instead, play opens us to a transformation of our everyday reality through its repeated practice because these worlds of play are *no less real* than the everydayness of survival. Fusing the real and playful worlds new horizons by which to configure flourishing crossings of the flesh open before me. These experiences create a resilience of awe-full hope so that I can adopt an attitude of wonder in the face of the anxiety and terror felt when disorienting experiences seem to threaten my everyday dealings with the world outside the intentional bounds of play.

Presence, wonder, and play form the backbone for thinking about what *I* do—how I might approach and engage the world—in order that *we* live-into the meaningful refractions of the *imago Dei* shaping this artful planet so that new modes of flourishing are made available. After all, as we previously affirmed, to *be* the image of

God is not something *I* am able to do. We live-into this *imago Dei* together; it is a pan-human capability for being-with-in-the-world particularly through our technobiogeochemical agency. No one of these features (presence, wonder, or play) forms a necessary or sufficient condition for being the *imago Dei*.

The consequence of this approach is that while *I* might carefully develop a disposition of presence, an attitude of wonder, and a consistent praxis of poietic play, the continued occurrence of the *imago Dei* might still fail; human being could very well go extinct before stable technobiogeochemical cycles take hold on the planet. After all, we have already stressed that it is not an inevitable destiny of the Anthropocene to become a Sapiezoic eon; it is not inevitable that human beings will be the ones who live-into this vision of the *imago Dei* as participants in transition from a living to an artful planet, even if it is the hopeful promise of our way of being-with-in-the-world. Living into presence, wonder, and play cannot guarantee the destiny of the *imago Dei* will be fulfilled, but exemplifying these qualities is the best that *I can do* to contribute to the continuing possibility of actualizing the *imago Dei*.

To complicate matters further, living into presence, wonder, and play is itself no easy task for the individual. None of these concepts provide a clear rule that could be directly instrumentalized and thereby delineate some universal path of behavior we might take to realizing the *imago Dei*. These concepts are highly contextual: general strategies that have to be constantly reapplied in the face of changing conditions. Moreover, a simple awareness of these concepts does not mean they can or will be immediately enacted as though their performance can be specifically commanded. We cannot simply say, "Play more!" and watch play take place. Nor can we command a sense of wonder ("Marvel now at the world!"). Finally, somehow "requiring" a sense of felt solidarity to the flesh as presence ("You must be present to others!") is downright nonsensical. These are all concepts that we must carefully and intentionally foster, even argue for, within our wider communities so that we provide the relaxed conditions in which play can take place and support the awareness of awe after terror that is wonder such that we realize our fundamental presence as being-with one another in the flesh.

It might be best to describe presence, wonder, and play as public aspirations. This disposition to presence in the flesh, attitude of wonder, and praxis of play are inspired by the specific sense of being the *imago Dei* envisioned in this work. They are the transreligious values inspired by this corroboration of the *imago Dei* and the Anthropocene in an astrobiological context.

Yet we might reverse the direction here as well. Could we call this vision of the *imago Dei* playful? My account of the *imago Dei* might encourage the development of presence, wonder, and play; however, the constructive framing of the *imago Dei* and the de-centering of typical sensibilities of how this doctrine applies to human individuals is itself, also, playful. More broadly, to connect with the focus of the previous chapter, not only the *imago Dei* but other doctrinal symbols could be thought of as playful ways of steeling ourselves toward or through disorientation—confronting it in awe-ful wonder—that then inspire a felt sense of presence and solidarity. If doctrinal symbols are playful means of wonderfully disorienting us so that we take up our presence in the flesh in new ways, then they can have an important role in public intellectual discourse to inspire a sense of moral revival and solidarity with anyone or anything that might be made present to us in the flesh.

Finally, it is important to note that though I have throughout this chapter always moved from presence through wonder to play, it is not my intention to indicate that these themes must always appear in some sort of serial order. This linearity is artificial. These concepts are part of an arc that describes a way of being-with-in-the-world wherein each concept is contemporaneous to the others. Each of the three concepts helps prop up the other two by doing something distinctive.

Play is what we can specifically practice. Presence and wonder are intentionalities fostered by playing, but the intentional disorientation of playing is something that we can actually *do*. Wonder is the only concept that is not explicitly indirect. Wonder is felt acutely in my direct experience of the everydayness of the world on account of the possible consequences of disorientation. By contrast, presence to the flesh describes a disposition toward realizing crossings in the flesh that intentionally disrupt the everydayness of our subjectivity, and play is a phenomenon that belongs neither to the subject who is playing nor the objects and spaces of play. It is understood explicitly in terms of practicing a certain intercorporeal confusion of the bodies, spaces, and objects involved in the playing so that each is only understood in terms of the immersive and temporary quality of the play-event. Finally, presence is the only one of the three concepts to deal with well-oriented experiences of the chiasm. Wonder and play operate based on disorientation—which is by definition meant to be impermanent. But presence describes an extension of these concepts; presence applies these ideas across the whole spectrum of possible crossings in the flesh.

So how can *we* be the image of God? What do *I* do in the midst of that image? We might each offer a promise. I will play more; I will linger in those moments of disorientation and wonder at the space between me and the world as we reach out toward it; I will dwell in the presence of the flesh and know that my existence matters. Confident in that assurance I will play even more; I'll invite others to feel the presence of the flesh by including them in my games— inviting them to find themselves as players in the alternative worlds that we build together. I will help them to see how they give meaning to me and others—how their exposure and their advance orients my place in the world. All of this I will do in the scope of a widened cosmos, with a deep concern for the fragility of our planetary flesh, wondering at the possibilities that open up when I accept that the *imago Dei* cannot be mine alone, but is the hope of human being as a refractive living-system participating in the unheralded adventing of an artful planet.

# Epilogue:
## *Ad Astra Per Aspera*

Usually translated for schoolchildren as "To the stars through diffi-culty," the Kansas State motto "Ad Astra Per Aspera" references the long process of Kansas achieving statehood. Yet it always indicated more than this alone. The Latin phrase *ad astra* hearkens back to passages from Virgil's *Aenead* and Seneca the Younger's *Hercules Furens* where it often refers to the difficulty of pursuing and achiev-ing any noble goal.

It is only after having lived in a small town on the Kansas prairies for a number of years now that I have begun to appreciate this state motto. Perhaps, though, my understanding is a bit more literal than originally intended. When I hear these words, I see the brilliance of a night sky that I can look up at with my children from our backyard and still catch wisps of the Milky Way. I can only imagine the bril-liance of what that night sky would have been when those words of the motto were first adopted—a time without the encroaching light pollution that now blankets the eastern United States.

It makes me sad to know how rare it is for people to see this nightly spectacle. We can still see the stars through science muse-ums and intentional trips to observatories. Still, we are losing some-thing from this hope of striving for the stars as our own incandes-cent glowing increasingly snuffs out a nightly brilliance. Moreover, the hope that can come from the disorienting wonder of seeing the stars—feeling just how tiny our own place in the universe truly is—is hardly something we can afford to lose today. If we let those stars realign what we think is possible, if we let them open us to queer new trajectories for re-orienting our place in the world, who knows how far we might go in doing the playful work of creating sustain-

188

able technobiogeochemical cycles while creating new senses of our felt presence in the flesh.

To close this work, I want to consider a few brief examples of addressing the environmental challenges of our Anthropocene age by invoking themes from the last two chapters: the three moods—presence, wonder, and play—and the concept of technobiogeochemical cycles. By no means is my intention for these examples to be exhaustive or systematic accounts of how we might approach developing ecological ethics correspondent to a broadened account of the *imago Dei*. They are examples meant to engage in the sort of play for which I am advocating. In short, the aim here is to put a slightly more practical face on what it looks like to open new possibilities for the flourishing of all creation. How do presence, wonder, and play operate as more than abstract values? What do we do to facilitate living into our planetary sense of the *imago Dei* as flesh? What does it pragmatically mean to facilitate technobiogeochemical cycles that resist runaway positive feedback? My hope is to make a case to that disgruntled student that "just philosophy" or "just theology" has a role to play in facilitating our capacity to imagine pragmatic solutions in the face of environmental crises.

### Desertification and Changes to Farming

The Sahara Desert is expanding. This is not new information; it is well known that most deserts expand and then contract with some regularity. However, a study published by Natalie Thomas and Sumant Nigam, which looked at the expansion of the Sahara Desert over the twentieth century, found southern expansion of the desert during the summers and northern expansion during the winters. This led to a 10 percent increase in size of the desert when defined in terms of annual rainfall. Importantly, the southern expansion of the desert reaches into the Sahel: a grassland region between the Sahara and the fertile, south savannas.[1] This finding is important not only for the Sahara, but for other subtropical deserts, where desertification affects the amount of arable land available for the production of food in regions that are often already food insecure.

Stopping expansion of the Sahara and other subtropical deserts is thus immensely important. For instance, expansion of the Gobi desert is such a problem that China's government has begun work on a new civic project: the Green Great Wall. It will include 88 million acres of forest planted along 3,000 miles of the Gobi Desert by 2050 in an effort to prevent the further desertification of arable land.[2] Recently, though, one team has made a significant new proposal regarding desertification in the journal *Science*. The significance of this new modeling should not be minimized even if the research team is quite clear that the predictive modeling is applicable only in the context of the Sahara. This plan might provide an imaginative spark for thinking about desertification in other subtropical desert regions as well.

The research team, led by Yan Li, built a model for an alternative Sahara Desert in which 20 percent is covered in solar panels and much of the rest is covered in wind turbines. Yan Li's research team predicts a significant increase in rain in the Sahel region (approximately 200–500 mm/year, which would have significant ecological impacts). The article is important because the environmental impact is not only related to the significant energy that would be produced by such a farm (four times as much as the planet currently consumes). The positive albedo-precipitation-vegetation feedback from the direct albedo effects of the solar panels and the decreased wind velocity due to friction from the turbines would help reclaim critical arable areas in the Sahel.[3]

The importance of this albedo effect should not be minimized. As Ning Zeng and Jinho Yoon have suggested, if the albedo effect of vegetation loss is factored into climactic models calculating anticipated desertification in subtropical regions, the predicted expansion of these deserts increases nearly 25 percent over standard models that do not account for this effect.[4] Moreover, if the plan to cover an area the size of the United States in solar panels and wind turbines sounds outlandish, consider the response of Eugenia Kalnay, a member of the research team doing the predictive modeling on the Sahara Desert energy farm, in popular reporting about this project.

> I told her that the whole scenario sounds like science fiction. Kalnay disagreed. "It would be science fiction if the technology was not available," she said.
> "So you could imagine it?"
> "Yes," she said, confidently.[5]

While combating desertification is certainly important, understanding how humans have contributed to the development and expansion of subtropical deserts has become increasingly important as well. A striking set of correlations have begun to develop between the spatial and temporal patterns of the initiation or intensification of agricultural and pastoralist practices with instances of desertification.[6] Keeping such a correlation in mind, we must both combat contemporary desertification and rethink various agricultural practices as they intensify strain on ecologically sensitive areas. This is particularly important in the area of the country where I live. According to the Kansas Department of Agriculture, farming accounts for 44.5 percent of the state's economic production and 88 percent of the state's land usage (over 46 million acres).[7]

I did not think too much about the farming processes by which my food was produced until I started living near farmers and agronomists in a state deeply tied to this industry. Certainly, I had a sense of issues related to production, distribution, and access to food, but I had no conception of the climate risks that are an inherent part of farming techniques until meeting a few farmers involved with a group called "No-Till on the Plains." This non-profit is advocating for farmers to implement a shift in technique: to stop tilling. Too often, we think of the soil used for farming as an infinite resource, but a longer view of history shows just how errant such a view would be. The techniques and crops that we so commonly farm on an annual basis strip the rich soil of arable lands.

The identified premise and purpose of No-Till on the Plains, however, goes beyond simply promoting environmentally safer practices: "It was decided that if man [*sic*] would examine the symbiotic relationships between the soil, soil properties, microbial life, plants, insects, nutrients and wildlife, he should be able to determine practical methods to eliminate the degradation of this, *his most valuable resource*."[8] What strikes me about the language of No Till on the Plains is its concern for what I have been calling the flourishing of intra-actions. It is the set of symbiotic relationships associated with the soil they want to foster and regenerate by avoiding any churning of the soil so that the complex and crucial interrelations of various bacterial microbes, fungi, and plant systems are not torn asunder.

Adopting "the most benign method of agriculture," No-Till on the Plains distances the practices for which they advocate from what they describe as a simplistic approach to sustainability. Recognizing that agriculture in almost all its forms has been—at least thus far in human history—a contributor to environmental degradation, they claim that we need solutions that go *beyond* sustainability.[9] By intentionally emulating existing natural cycles that foster the symbiotic relationships crucial to soil health, No-Till on the Plains seeks to educate farmers about developing practices and technologies that will regenerate and enrich soil health in the future.

Other groups in Kansas, such as the Land Institute, would go a crucial step further than No-Till on the Plains. Beyond emulating existing natural processes, they are also seeking to develop perennial grain and seed crops that can be integrated into a more ecologically stable system of intensified polycultures. These perennial grain and seed crops would not be subject to the same cycle of disturbance that dictates the planting and harvesting of annual, monocultured crops (disturbances that, though lessened, remain inherent to the use of annual crops and "cover cropping" in no-till agricultural practices).[10]

Because perennial crops do not have to be reseeded every year, they provide a more robust root structure for soil regeneration, fostering symbiotic relationships crucial for soil health, increasing water infiltration of the soil, and increasing soil carbon sequestration (thus decreasing atmospheric carbon dioxide over time). By intentionally planting these perennial grain and seed crops with respect for biodiversity, researchers at the Land Institute believe that farming could actually have regenerative impacts at the level of whole ecosystems.[11] Moreover, combining various "climate smart agricultural practices" (such as those advocated for by No-Till on the Plains and the Land Institute) has been shown to not only decrease environmental footprints but also *increase yields and profitability* in agricultural settings outside the United States.[12]

In light of the schemas that I have laid out in the final two chapters, I want to suggest that the scientific initiatives described here (whether in terms of agronomy or desertification) are forms of *play*. Here, the scientific imagination attempts to construe rules for how we might behave. Scientists play at imagining a new world of energy and agricultural practices out of step with the realities dictated by current understandings. In particular, all of these forms of play concern how the use of technology does or does not integrate with wider biogeochemical cycles.

These are forms of play that imagine what a human living-system understood as a particular type of technobiogeochemical intra-action might actually entail in terms of practical changes in various contexts. This variety of contexts, issues, and solutions also indicates that these playful ways of becoming the *imago Dei* are primarily instantaneous rather than planetary (as distinguished in Chapter 7). The specificity entailed by playing means that it will always provide ways of thinking about what it means to be the *imago Dei* at an instantaneous scale (what the prospects of a particular sort of technological intra-action might entail). To work toward understanding the implications of being the *imago Dei* at a truly planetary scale requires engaging features of wonder and presence to which we must now turn, but which also take us back to the work of the Land Institute.

### Consciousness Raising and Climate Refugees

If the work of the Land Institute were only concerned with ecological intensification and the development of perennial grain and seed crops, I would describe the work they do as a sort of play. Since 2015, however, they have also been engaged in developing a course of education they call "Ecosphere Studies." In their own words,

> The ecosphere is the creative, living globe that is our home. Rather than viewing Earth as a planet composed of living and non-living parts, and people as separate from our environments, Ecosphere Studies understands human communities as nestled within ecosystems.
>
> We are part of an intricately interdependent system, from which came an emergent property of life. This perspective allows us to address enduring questions of identity and ethics in new ways.[13]

The aim here is broader than just the discovery of scientific knowledge or engagement in a form of play that imagines new possibilities that challenge the boundaries of what constitute reality. Here there is an intentional effort at cultural transformation and consciousness raising that seeks to address how mainstream educational systems must shift if they can no longer rely on an economy driven by extractive agricultural practices. In the language that I have introduced here, Ecosphere Studies represents an intentional effort to disorient and reorient the sense we have for what is intra-actively possible. More than play, here the goal is to inspire wonder.

Thinking about the matrix of orientation-toward and orientation-around developed in the previous chapter, we might say that Ecosphere Studies is intended to overturn our everyday sense of a distancing from the intricate interdependence with wider ecological systems reflecting a coordinate misjudgment regarding what features of those systems are actually proximate to us. It represents a process for disorienting us and re-orienting us in realizing our proximity to these ecological interdependencies. This is about fostering an attitude of wonder that makes us more receptive to the imaginative possibilities pursued through play.

Various religiously based initiatives that seek to inspire awareness about ecological devastation or inspire action within specific religious communities also tend to be driven by this aim of inspiring an attitude of wonder. Take, for instance, a recent initiative within the ELCA: Lutherans Restoring Creation (LRC).

This grassroots initiative works to engage various congregations and other units of this church body in connecting their social and theological affirmations with practices of caring for the environment. Citing a long history of Lutheran concern for creation care, this group intentionally seeks to empower existing institutions to face the challenges of environmental degradation in new ways.[14] For instance, the local LRC group for Kansas and Missouri has been hosting creation care workshops and providing congregations with a self-organizing kit. Grounded in a set of common theological tenets, the kit encourages congregations to approve a mission statement, form an action plan and "green team" to carry it out, promote this concern for creation care to wider publics, and then revise the guiding action plan on an annual basis.

Many of the themes and instructions in the kit are familiar to anyone who has worked for environmental justice before; some suggestions are quite different though. For instance, the kit has a set of suggestions for thinking of "The church as an alternative community" that includes instructions to,

> Know your property as an "Earth Community." Get to know the trees, plants, animals, insects, birds, and other creatures who live with you on this space. Live in such a way that all of you may thrive together. Pray for them. Worship with them. Include some in your church directory as your creation family.[15]

While I have yet to see the greater prairie chicken included in the directories of various congregations in the area, the intent here is clear. Elements of the wider "earth community" that were not considered to be proximate to a "worshipping community" are being invoked here. This would be disorienting; just imagine looking for "Pryor" in a directory and instead of finding a person you see "Prairie Chicken." The stakes in this example are small, but there is an effort here intentionally to disorient.

We could invoke the language introduced in the last chapter concerning wonder to describe the effects of seeing other members of the "Earth Community" appear in a church directory. We immediately sense tension between a desire to contact members of the congregation (a subjective orientation-toward) and the horizon of who or what is included in the directory (an objective orientation-around). The, albeit minimal, sense of anxiety produced by this disjunction of flesh could cause abandonment of the members of the wider Earth community included in the directory. We can imagine a church member shaking his head, muttering to himself about this silly entry of "Prairie Chicken," and immediately recontextualizing his experience of encountering this entry instead of "Pryor" in light of the wider work of this "Green Team" in his congregation. Alternatively, this disjunction could lead to a new opening in the flesh. The church member remains with the minimal sense of anxiety and produces a double movement that responds with an awe for the breadth of the wider intra-actions of the Earth community.

While a project like that of Lutherans Restoring Creation works through intentionally targeted communities held together theologically, and the Ecosphere Studies of the Land Institute aims more broadly at reforming mainstream education, both focus on facilitating the development of wonder through a kind of consciousness raising that leads to a felt solidarity with our wider ecological interdependencies. These efforts explore the hidden assumptions of what constitute the normative intra-actions that are immediately recognizable and the ways in which this recognizability is a limit on other forms of intra-action that appear in disorienting these inherited preconceptions. This disorientation encourages a reorientation that makes the excised intra-actions proximate and available in ways that had previously been impossible and, I would suggest, is intended to form a sense of identity that is rooted in the alterity that is constitutive of our planetary being. Wonder should cause us to see the world differently; each of these projects is seeking to make seemingly invisible or distant intra-actions proximate to us in new ways.

Yet this remains an intermediate step in realizing the possibilities of our human living-system becoming the *imago Dei*. We must work to move beyond wonder toward presence, to move beyond re-orientation to newfound senses of widened responsibility. For instance, I recently heard the ELCA Bishop of Alaska, Shelley Wick-strom, describe Shishmaref as a poster child for climate refugees.[16] Located on Serichef Island, this village of approximately 140 homes is built on a stretch of sand dunes 120 miles northwest of Nome. "Remote" only begins to describe Shishmaref, but the challenges the residents face in a twenty-first-century global economy magnify a trend that we find in other remote rural communities. Few homes have running water, groceries are unbelievably expensive (a gallon of milk is twelve dollars), and an increasing percentage of adults in the community are no longer in the cash labor force (approximately 46 percent).

What is perhaps most notable about Shismaref, though, is the ef-fect of climate change on this community and the age of its aver-age resident. It is expected that one in nine people will be climate refugees by the year 2050, though we should expect this ratio could be far grimmer in Artic areas where air temperature changes are double that of what we find in more temperate areas. Moreover, in Shishmaref, the average age is 22. Half of the residents are younger than 18; these are people who should easily live to see the devastat-ing shifts that climate experts anticipate to be upon us by the mid-twenty-first century.

Historically, these Alaskan Natives, who are largely Inupiat, were migratory people; they settled after the establishment of a school in 1901 in an area that was a good place to hunt seal and fish during the spring. Not incorporated until 1969, census data is recorded back as far as 1920. It is reasonable to say the people of Shishmaref have been in this place for the past century.

Nonetheless, with decreasing sea ice, the coasts of this narrow island are becoming increasingly susceptible to erosion—so much so that key pieces of infrastructure (such as the road next to the air-strip, which is the only way the village could be evacuated) are be-ing affected. Seeing these impending devastating effects, the people voted to move as early as 2002 and again voted in 2016 to move to one of three federally designated sites. It was not until 2015, though, that an economic-development agency (the Denali Com-mission) was designated to coordinate federal funding for Alaskan communities threatened by climate change; that designation, how-ever, came without any allocated budget. More recently, in 2016, President Trump's proposed federal budget sought to eliminate the Denali Commission entirely, though after protests erupted, funding was found to avoid such a calamity.

Even with a slim majority willing to move the village (the 2016 vote was 94–78 in favor of moving), the three designated sites for Shismaref are devoid of infrastructure. Even if the 2 million dollars proposed for relocation were to be allocated, it would take significant time to develop the sites so that they might be accessed and habitable. To make matters worse, because the village has voted to move, they no longer have access to typical state funding for existing infrastructure improvements.

What struck me about Bishop Wickstrom's reflections on the Inupiat of Shishmaref was her engagement with the subsistence culture of these arctic islanders. Describing this as a culture that is "more I-Thou than I-It," invoking Martin Buber's rich ontology, she notes this community lives out the *imago Dei*. Describing a mentality driven by kinship, compassion, and respect for one another and nature, she relayed a short anecdote:

> When a fish is caught or a seal is killed, the hunter/fisher thanks the animal for giving its life to feed the hunter's family. When an animal crosses your path, you, as a hunter, are obligated to kill it to honor it. I've never heard a hunter talk about getting "my caribou." It is always "This quarter for Grandparents; this quarter for auntie; this quarter for my neighbor; this quarter for the freezer."
>
> This is not animism where a property of a person is projected onto the surroundings. This is *imago Dei* in creation. . . . In a subsistence culture, you are of the earth, of the sea, not on it.[17]

Here, the mutuality of existence takes precedence in the understanding of traditional Inupiat values as these intersect with the understanding of the injunction to be stewards for creation. How should we respond to a culture that has stressed a sense of mutualism in the *imago Dei*, one that I have advocated for here, while also bearing the brunt of our unwieldy entry into the Anthropocene? Surely the relocation of these climate refugees will irrevocably disrupt a way of life that has been honed over the better part of a century.

The people of Shishmaref and other climate refugees illustrate a challenge inherent to my approach. What do we do when the field of play—the free space to imagine alternative realities—shrinks? The plight of these people is something that we can only sufficiently address if we have developed a disposition for presence in the flesh that seeks to hold the intra-action of playful farming practices in Kansas and wondrous attitudes that we begin to form in finding Prairie Chickens in church directories with the devastating effects of rising sea levels on an island that seems worlds apart from these land locked locales. The plight of the people of Shishmaref is my plight if I hope to refract new ways of meaningfully being-together in the midst of creation as the *imago Dei*. In presence, I understand myself to be part of a crossed-phenomenon with the people of Shishmaref, a phenomenon that is more basic even than the parts into which it might be subdivided. If we can do this, then in presence we must make ourselves into a space that vocatively declares to the Inupiat, "Here I am." We must make ourselves and our resources into a space for their flourishing—being present to them even in the midst of discerning what flourishing might mean in the new and difficult situations the people of Shishmaref must confront. It is hard to imagine how we do this without an astrobiological perspective that inextricably situates living-systems into the contexts of their habitable environments with a deep appreciation for the fragility and precarity of those planetary locales.

The examples offered here are intentionally diverse. They move from specific efforts of research scientists to conceptualize tipping points in the climate history of the Holocene; to the work of agronomists and farmers developing sustainable agricultural praxes; to initiatives for guiding local congregations in reflection on their environmental footprint; to the real concern for how we might best respond to climate refugees who have modeled a mutualism with the natural world we might advocate for as part of the *imago Dei*, even as they are driven from their homes by the choices of many others. Each of these examples is speaking to issues that relate to what it might practically mean for us to live into a planetary vision of the *imago Dei*.

What happens when the Anthropocene and the *imago Dei* become corroborative symbols in the astrobiological contexts that shape our engagement with the world today? My argument has been that, in the face of various instances of ecological crises, the Anthropocene symbolizes the existential concerns at stake in this devastation so that we better understand that our way of meaningfully orienting our existence toward the natural world is askew. To remember that we are the *imago Dei* can give us courage to stay with the trouble of this disorientation a moment longer and imaginatively play out new realities that confront the inevitable ecological devastations that have been wrought upon the earth. Perhaps it can even help us develop a disposition of presence: imagining how we might affirm *techno*biogeochemical cycles that reintegrate our human ways of being in the world into the long honed intra-actions of the living-systems and habitable environments that constitute the particular precarity of this living planet. Perhaps we can still become an artful planet, refracting new possibilities of meaningful existence for the lives of tiny terrestrial microbes or any number of the other voice-less features of creation so that if we find one day we are also living with tiny aliens, we will be better prepared to become an artful universe.

# Acknowledgments

Writing a book is always the work of many people, too many people to possibly thank them all. I am immensely grateful to the team at Fordham University Press, especially Thomas Lay, and the editors of the *Groundworks* series, Forrest Clingerman and Brian Treanor. Their suggestions and guidance enhanced this project tremendously.

My colleagues provided continuous support throughout the long process of writing. Without William Storrar's leadership and the generous support of the Center of Theological Inquiry to study the societal implications of astrobiology, I would not have been able to write this book. My colleagues in Princeton at the Center shaped my thinking in ways I could not possibly have anticipated. I am especially grateful to Andrew Davison, John McCarthy, and Olli-Pekka Vainio, all of whom offered valuable input at different points during our residency together. My colleagues at Bethany have continued to support my work on this project in myriad ways. I will forever be grateful to Robert Carlson who, as Provost, encouraged me to take a year away from the campus to be at the Center of Theological Inquiry. Tyler Atkinson, Arminta Fox, Mary Beth Harris, Marcus Hensel, Greg LeGault, Rebecca Miller, and Kristin Van Tassel have listened tirelessly to my meandering thoughts, pushed me to be more clear when communicating esoteric concepts, and generously and unconditionally supported not only me but my family as I finished this project.

Finally, and most important, I am grateful to my wife, Rachael. In the swirl of changes that have come since I began this project, she has been steadily present for me, always wondering with me about the arguments I am making, and playfully (but sternly) reminding me to take a break and be part of the world I hope we live into for our children instead of just writing about it.

# Notes

**Introduction: Being in Outer Space**

1. Rebecca Whittaker, "Learn About Spacesuits," NASA, June 5, 2013, http://www.nasa.gov/audience/foreducators/spacesuits/home/clickable_suit_nf.html.

2. Carl Sagan et al., "A Search for Life on Earth from the Galileo Spacecraft," *Nature* 365, no. 6448 (October 21, 1993): 716.

3. Sagan, "A Search for Life," 719–20.

4. Sagan, "A Search for Life," 720.

5. Sagan, "A Search for Life," 720.

6. Ewert Cousins, *Bonaventure and the Coincidence of Opposites* (Chicago: Franciscan Herald Press, 1978), 25.

7. David Nikkel, "Embodying Ultimate Concern," in *The Body and Ultimate Concern: Reflections on an Embodied Theology of Paul Tillich*, ed. Adam Pryor and Devan Stahl (Macon, Ga.: Mercer University Press, 2018), 34.

8. Carl Pilcher, "Questing for Life: The Scientific Challenge" (Center of Theological Inquiry Winter Symposium 2017, Princeton, N.J., January 30, 2017).

9. See Mark Lewis Taylor, *The Theological and the Political* (Minneapolis: Fortress Press, 2011), chap. 1.

10. Robert Cummings Neville, *Ultimates: Philosophical Theology*, vol. 1 (Albany: SUNY Press, 2013), 15.

11. Paul Tillich, "The Effects of Space Exploration on Man's Condition and Stature," in *The Future of Religions*, ed. Jerald C. Brauer (New York: Harper & Row, 1966), 39–51.

12. Catastrophic change is used here in the technical sense, meaning any—positive or negative—dramatic change in the state of the planet.

13. Edward Farley, *Deep Symbols: Their Postmodern Effacement and Reclamation* (Valley Forge, Pa.: Trinity Press International, 1996), 23. Emphasis mine.

14. See Mark C. Taylor, "End the University As We Know It," *The New York Times*, April 27, 2009, sec. Opinion, http://www.nytimes.com/2009/04/27/opinion/27taylor.html.

15. P. L. Rosenfield, "The Potential of Transdisciplinary Research for Sustaining and Extending Linkages between the Health and Social Sciences," *Social Science & Medicine* 35, no. 11 (December 1992): 1343–57.

16. J. T. Klein, *Crossing Boundaries: Knowledge, Disciplines, and Interdisciplinarities* (Charlottesville: University of Virginia Press, 1996).

17. Daniel Stokols, "Toward a Science of Transdisciplinary Action Research," *American Journal of Community Psychology* 38, nos. 1–2 (2006): 67.

18. A. K. Giri, "The Calling of a Creative Transdisciplinarity," *Futures* 34, no. 1 (2002): 103.

19. See T. Horlick-Jones and J. Sime, "Living on the Border: Knowledge, Risk and Transdisciplinarity," *Futures* 36, no. 4 (2004): 441–56; M. Lengwiler et al., "Between Charisma and Heuristics: Four Styles of Interdisciplinarity," *Science & Public Policy* 33, no. 6 (2006): 423–34; and Gertrude Hirsch Hadorn et al., "Implications of Transdisciplinarity for Sustainability Research," *Ecological Economics* 60, no. 1 (2006): 119–28.

20. Robin W. Lovin et al., "Introduction: Theology as Interdisciplinary Inquiry—The Virtues of Humility and Hope," in *Theology as Interdisciplinary Inquiry: Learning with and from the Natural and Human Sciences*, ed. Robin W. Lovin and Joshua Mauldin (Grand Rapids: W. B. Eerdmans, 2017), xxiii–xxiv.

21. See Lindsay E. Hays, ed., "Astrobiology Strategy" (NASA, October 2015), https://astrobiology.nasa.gov/uploads/filer_public/01/28/01283266-e401-4dcb-8e05-3918b21edb79/nasa_astrobiology_strategy_2015_151008.pdf.

22. My work remains deeply indebted to Robert Russell's model of Creative Mutual Interaction, which recognizes the importance of creating analytical clarity in order to facilitate cross-disciplinary conversation with multiple (and sometime competing) aims. See *Cosmology—from Alpha to Omega: The Creative Mutual Interaction of Theology and Science* (Minneapolis: Fortress Press, 2008).

23. Celia Deane-Drummond "Looking at Humans through the Lens of Deep History," in *Theology as Interdisciplinary Inquiry: Learning with and from the Natural and Human Sciences*, ed. Robin W. Lovin and Joshua Mauldin (Grand Rapids: W. B. Eerdmans, 2017), 11–15.

24. Wentzel van Huyssteen, *The Shaping of Rationality: Toward Interdisciplinarity in Theology and Science* (Grand Rapids: W. B. Eerdmans, 1999); or Wentzel van Huyssteen, *Alone in the World?: Human Unique-*

*ness in Science and Theology* (Grand Rapids: W. B. Eerdmans, 2006), chap. 1.

25. See Andrew Linzey, *Animal Theology* (Urbana: University of Illinois Press, 1995); Celia E. Deane-Drummond and David Clough, *Creaturely Theology: God, Humans and Other Animals* (London: Hymns Ancient & Modern Ltd, 2009); David L. Clough, *On Animals: Volume I: Systematic Theology* (Edinburgh: T & T Clark, 2014).

26. Max L Stackhouse, "Civil Religion, Political Theology and Public Theology: What's the Difference?," *Political Theology* 5, no. 3 (July 2004): 275–93; John Thatamanil, "Transreligious Theology as the Quest for Interreligious Wisdom: Open Theology," *Open Theology* 2, no. 1 (May 27, 2016): 354–62.

## 1. Exoplanets and Icy Moons and Mars, Oh My!

1. See C. R. Woese, O. Kandler, and M. L. Wheelis, "Towards a Natural System of Organisms: Proposal for the Domains Archaea, Bacteria, and Eucarya," *Proceedings of the National Academy of Sciences of the United States of America* 87, no. 12 (June 1990): 4576–79.

2. A. Wolszczan and D. A. Frail, "A Planetary System around the Millisecond Pulsar PSR1257 + 12," *Nature* 355, no. 6356 (January 9, 1992): 145–47.

3. Bruce Jakosky, *Science, Society, and the Search for Life in the Universe* (Tucson: University of Arizona Press, 2006), 46–48.

4. Lindsay E. Hays, ed., "Astrobiology Strategy" (NASA, October 2015), https://astrobiology.nasa.gov/uploads/filer_public/01/28/01283266 -e401-4dcb-8e05-3918b21edb79/nasa_astrobiology_strategy_2015_151008 .pdf.

5. Tahir Yaqoob, *Exoplanets and Alien Solar Systems* (Baltimore: New Earth Labs, 2011), chap. 3.

6. Researchers call this being "edge-on." Statistically only one to two percent of exoplanetary systems orbiting their star would be visible to us by the transit method, which makes the sheer number of planets that Kepler has detected truly remarkable. See Erik A. Petigura, Andrew W. Howard, and Geoffrey W. Marcy, "Prevalence of Earth-Size Planets Orbiting Sun-like Stars," *Proceedings of the National Academy of Sciences* 110, no. 48 (November 26, 2013): 19273–78.

7. R. K. Kopparapu et al., "Habitable Zones around Main-Sequence Stars: New Estimates," *Astrophysical Journal* 765, no. 2 (March 10, 2013): 131. Astrobiologists sometimes make a distinction between the conservative habitable zone, based on radiative-convective climate modeling, and the optimistic habitable zone, which takes into account the additional possibility that Mars and Venus may have each had liquid water on their surfaces previously.

8. See Jon M. Jenkins, et al., "Discovery and Validation of Kepler-452b: A 1.6 R$_⊕$ Super Earth Exoplanet in the Habitable Zone of a G2 Star," *The Astronomical Journal* 150, no. 2 (2015): 56; J. F. Kasting, D. P. Whitmire, and R. T. Reynolds, "Habitable Zones around Main Sequence Stars," *Icarus* 101 (1993): 108–28.

9. Jon Austin, "Kepler-452b: How Long Would It Take Humans to Reach 'Earth 2' and Could We Live There?," *Express.co.uk*, July 28, 2015, http://www.express.co.uk/news/science/594133/Kepler-452b-How-long-take-humans-reach-Earth-2-could-we-live-there.

10. Jingjing Chen and David Kipping, "Probabilistic Forecasting of the Masses and Radii of Other Worlds," *The Astrophysical Journal* 834, no. 1 (2017): 17.

11. See John Wenz, "Study Casts Doubt on Existence of a Potential 'Earth 2.0,'" *Scientific American*, April 9, 2018, https://www.scientific american.com/article/study-casts-doubt-on-existence-of-a-potential-earth -2-0/. See Fergal Mullally et al., "Kepler's Earth-like Planets Should Not Be Confirmed Without Independent Detection: The Case of Kepler-452b," *The Astronomical Journal* 155, no. 5 (April 25, 2018): 210.

12. E. F. Guinan and N. D. Morgan, "Proxima Centauri: Rotation, Chromospheric Activity, and Flares," *Bulletin of the American Astronomical Society* 28 (May 1996): 942.

13. The mass cannot be given as a determinate value on this method. One would have to know the angle of inclination of the planetary system in relation to our own to make such a calculation. The mass given is a minimum because it could be that we are not seeing the full wobble of a given star. See Yaqoob, *Exoplanets and Alien Solar Systems*, chap. 3; and David C. Catling, *Astrobiology: A Very Short Introduction* (Oxford: Oxford University Press, 2013), 111–12.

14. Guillem Anglada-Escudé et al., "A Terrestrial Planet Candidate in a Temperate Orbit around Proxima Centauri," *Nature* 536, no. 7617 (August 25, 2016): 437–40. Placement of Proxima Centauri b within the conservative habitable zone is made in reference to the estimates from Kopparapu et al., "Habitable Zones Around Main-Sequence Stars," 131. As much as possible I will avoid the term "red dwarf" because some M-type stars are *not* red dwarfs and sometimes "red dwarf" designates both M-type and K-type stars. What matters here is that most main-sequence stars we know of are M-type and there are some distinctive issues regarding the habitability of exoplanets around these stars. See Martin J. Heath et al., "Habitability of Planets Around Red Dwarf Stars," *Origins of Life and Evolution of the Biosphere* 29, no. 4 (August 1999): 405–24.

15. David M. Kipping et al., "No Conclusive Evidence for Transits of Proxima B in MOST Photometry," *The Astronomical Journal* 153, no. 3 (February 2, 2017): 93.

16. Rory Barnes et al., "The Habitability of Proxima Centauri b I: Evolutionary Scenarios," *arXiv:1608.06919 [Astro-Ph]*, August 24, 2016, http://arxiv.org/abs/1608.06919.

17. Rodrigo Luger and Rory Barnes, "Extreme Water Loss and Abiotic $O_2$ Buildup on Planets Throughout the Habitable Zones of M Dwarfs," *Astrobiology* 15, no. 2 (January 28, 2015): 119–43.

18. Sean N. Raymond, Rory Barnes, and Avi M. Mandell, "Observable Consequences of Planet Formation Models in Systems with Close-in Terrestrial Planets," *Monthly Notices of the Royal Astronomical Society* 384 (February 1, 2008): 663–74.

19. Dimitra Atri, "Modelling Stellar Proton Event-Induced Particle Radiation Dose on Close-in Exoplanets," *Monthly Notices of the Royal Astronomical Society: Letters* 465, no. 1 (February 11, 2017): L34–38; Vladimir S. Airapetian et al., "How Hospitable Are Space Weather Affected Habitable Zones? The Role of Ion Escape," *The Astrophysical Journal Letters* 836, no. 1 (2017): L3. More recently, the announcement of Ross 128-b at eleven light years away has emphasized many of the same points found in the announcement of Proxima-b, but the M-class star Ross 128 flares far less. See X. Bonfils et al., "A Temperate Exo-Earth around a Quiet M Dwarf at 3.4 Parsecs," *Astronomy & Astrophysics*, November 15, 2017.

20. Peter E. Driscoll and Rory Barnes, "Tidal Heating of Earth-like Exoplanets around M Stars: Thermal, Magnetic, and Orbital Evolutions," *Astrobiology* 15, no. 9 (September 1, 2015): 739–60. This gravitational pull causing tidal heating may, hypothetically, also occur in relation to the presence of other planets in the system depending on orbit and size. Importantly, there is a yet unconfirmed possibility that a Proxima Centauri c may exist with an orbital period somewhere between 60 and 500 days, but stellar activity and inadequate sampling have prevented its confirmation. See Anglada-Escudé et al., "A Terrestrial Planet Candidate," 439.

21. However, Proxima Centauri is gravitationally bound to the Alpha Centauri star system. Alpha Centauri A and B (which appear as one star in the night sky to the naked eye) are far more tightly bound than Proxima Centauri, which circles this star system approximately once every 500,000 years. If an exoplanet were found around either Alpha Centauri A or Alpha Centauri B, then that exoplanet would be the closest to the Earth while Proxima Centauri was transiting on the far side of this star system. Eventually, Proxima Centauri b would again be our closest exoplanet neighbor when Proxima Centauri returned to being on the near side of this star system. See J. G. Wertheimer and G. Laughlin, "Are Proxima and Alpha Centauri Gravitationally Bound?," *The Astronomical Journal* 132, no. 5 (2006): 1995.

22. See Kenneth Chang, "One Star Over, a Planet That Might Be Another Earth," *The New York Times*, August 24, 2016, https://www.nytimes

.com/2016/08/25/science/earth-planet-proxima-centauri.html; Megan Gannon, "Aliens Next Door: Does Proxima B Host Life?," Space.com, August 24, 2016, http://www.space.com/33846-proxima-b-alien-life-hunt .html; Mike Wall, "Found! Potentially Earth-Like Planet at Proxima Centauri Is Closest Ever," Space.com, August 24, 2016, http://www.space .com/33834-discovery-of-planet-proxima-b.html; or Alexandra Witze, "Earth-Sized Planet around Nearby Star Is Astronomy Dream Come True," *Nature News* 536, no. 7617 (August 25, 2016): 381.

23. P. Lubin, "A Roadmap to Interstellar Flight," *Journal of the British Interplanetary Society* 69 (2016): 40–72; and Gabriel Popkin, "What It Would Take to Reach the Stars," *Nature News* 542, no. 7639 (February 2, 2017): 20.

24. Lin Edwards, "IKAROS Unfurls First Ever Solar Sail in Space," *PhysOrg.com*, June 11, 2010, https://phys.org/news/2010-06-ikaros-unfurls -solar-space.html.

25. Michaël Gillon et al., "Seven Temperate Terrestrial Planets around the Nearby Ultracool Dwarf Star TRAPPIST-1," *Nature* 542, no. 7642 (February 23, 2017): 456–60; Sarah Ballard, "Predicted Number, Multiplicity, and Orbital Dynamics of TESS M Dwarf Exoplanets," *ArXiv:1801.04949 [Astro-Ph]*, January 15, 2018.

26. Rory Barnes, "Opportunities and Obstacles for Life on Proxima B," *Pale Red Dot*, August 28, 2016, https://palereddot.org/opportunities-and -obstacles-for-life-on-proxima-b/.

27. Perhaps like Gwyneth Jones's science-fiction novel *Proof of Concept* (Tor.com, 2017).

28. Craig O'Neill and Francis Nimmo, "The Role of Episodic Overturn in Generating the Surface Geology and Heat Flow on Enceladus," *Nature Geoscience* 3, no. 2 (February 2010): 88–91; and John R. Spencer and Francis Nimmo, "Enceladus: An Active Ice World in the Saturn System," *Annual Review of Earth and Planetary Sciences* 41, no. 1 (2013): 693–717.

29. The plumes make Enceladus a good venue for continued study. See Shannon M. MacKenzie et al., "THEO Concept Mission: Testing the Habitability of Enceladus's Ocean," *Advances in Space Research* 58, no. 6 (September 2016): 1118.

30. Robert H. Brown et al., "Composition and Physical Properties of Enceladus' Surface," *Science* 311, no. 5766 (March 10, 2006): 1425–28; Mikhail Y. Zolotov, "An Oceanic Composition on Early and Today's Enceladus," *Geophysical Research Letters* 34, no. 23 (December 16, 2007): L23203; D. Alex Patthoff and Simon A. Kattenhorn, "A Fracture History on Enceladus Provides Evidence for a Global Ocean," *Geophysical Research Letters* 38, no. 18 (September 28, 2011): L18201; F. Postberg et al., "A Salt-Water Reservoir as the Source of a Compositionally Stratified Plume on Enceladus," *Nature* 474, no. 7353 (June 30, 2011): 620–22.

31. G. Tobie, O. Čadek, and C. Sotin, "Solid Tidal Friction above a Liquid Water Reservoir as the Origin of the South Pole Hotspot on Enceladus," *Icarus* 196 (August 1, 2008): 642–52.

32. Nicolas Rambaux et al., "Librational Response of Enceladus," *Geophysical Research Letters* 37, no. 4 (February 1, 2010): L04202; L. Iess et al., "The Gravity Field and Interior Structure of Enceladus," *Science* 344, no. 6179 (April 4, 2014): 78–80; and P. C. Thomas et al., "Enceladus's Measured Physical Libration Requires a Global Subsurface Ocean," *Icarus* 264 (January 2016): 37–47.

33. M. M. Hedman et al., "An Observed Correlation between Plume Activity and Tidal Stresses on Enceladus," *Nature* 500, no. 7461 (August 8, 2013): 182–84.

34. J. Hunter Waite et al., "Cassini Ion and Neutral Mass Spectrometer: Enceladus Plume Composition and Structure," *Science* 311, no. 5766 (March 10, 2006): 1419–22; J. H. Waite, Jr. et al., "Liquid Water on Enceladus from Observations of Ammonia and $^{40}$Ar in the Plume," *Nature* 460, no. 7254 (July 23, 2009): 487–90.

35. M. E. Perry et al., "Cassini INMS Measurements of Enceladus Plume Density," *Icarus* 257 (September 1, 2015): 139–62.

36. See Christopher R. Glein, John A. Baross, and J. Hunter Waite Jr., "The pH of Enceladus' Ocean," *Geochimica et Cosmochimica Acta* 162 (August 1, 2015): 202–19. A key to this line of thinking is that we should find native hydrogen gas in the plume if serpentinization is occurring. Not confirmed, the evidence does seem highly suggestive. See J. Hunter Waite et al., "Cassini Finds Molecular Hydrogen in the Enceladus Plume: Evidence for Hydrothermal Processes," *Science* 356, no. 6334 (April 14, 2017): 155–59.

My focus has been on Enceladus, but wider analogues between extremophiles and potential lifeforms on icy moons exist. See Andrew Martin and Andrew McMinn, "Sea Ice, Extremophiles and Life on Extra-Terrestrial Ocean Worlds," *International Journal of Astrobiology* 17, no. 1 (January 2018): 1–16.

37. Deborah S. Kelley et al., "A Serpentinite-Hosted Ecosystem: The Lost City Hydrothermal Field," *Science* 307, no. 5714 (March 4, 2005): 1428–34; Christopher P. McKay et al., "The Possible Origin and Persistence of Life on Enceladus and Detection of Biomarkers in the Plume," *Astrobiology* 8, no. 5 (October 2008): 909–19; Hsiang-Wen Hsu et al., "Ongoing Hydrothermal Activities within Enceladus," *Nature* 519, no. 7542 (March 12, 2015): 207–10; N. G. Holm et al., "Serpentinization and the Formation of $H_2$ and $CH_4$ on Celestial Bodies (Planets, Moons, Comets)," *Astrobiology* 15, no. 7 (July 1, 2015): 587–600. Notably, there is not consensus that serpentinization is a sufficient energy source for the emergence of life. See Robert Pascal, "Physicochemical Requirements Inferred for Chemical Self-Organization Hardly Support an Emergence of

Life in the Deep Oceans of Icy Moons," *Astrobiology* 16, no. 5 (April 26, 2016): 328–34.

38. William Martin and Michael J. Russell, "On the Origins of Cells: A Hypothesis for the Evolutionary Transitions from Abiotic Geochemistry to Chemoautotrophic Prokaryotes, and from Prokaryotes to Nucleated Cells," *Philosophical Transactions of the Royal Society B: Biological Sciences* 358, no. 1429 (January 29, 2003): 59–85; William Martin et al., "Hydrothermal Vents and the Origin of Life," *Nature Reviews Microbiology* 6, no. 11 (November 2008): 805–14.

39. For instance, see Mike Wall, "Life-Hunting Mission Would Bring Samples Back from Saturn Moon Enceladus," Space.com, September 21, 2015, http://www.space.com/30598-saturn-moon-enceladus-sample -return-mission.html. For a more modest mission proposal, see MacKenzie et al., "THEO Concept Mission," 1117–37.

40. Xianzhe Jia et al., "Evidence of a Plume on Europa from Galileo Magnetic and Plasma Wave Signatures," *Nature Astronomy*, May 14, 2018, 1.

41. See William K. Hartmann and Gerhard Neukum, "Cratering Chronology and the Evolution of Mars," *Space Science Reviews* 96 (2001): 165–94; and Jean-Pierre Bibring et al., "Global Mineralogical and Aqueous Mars History Derived from OMEGA/Mars Express Data," *Science* 312, no. 5772 (April 21, 2006): 400–4.

42. J. Carter et al., "Hydrous Minerals on Mars as Seen by the CRISM and OMEGA Imaging Spectrometers: Updated Global View," *Journal of Geophysical Research: Planets* 118, no. 4 (April 1, 2013): 831–58; S. W. Squyres and A. H. Knoll, *Sedimentary Geology at Meridiani Planum, Mars* (Amsterdam: Elsevier, 2005); and James J. Wray et al., "Orbital Evidence for More Widespread Carbonate-Bearing Rocks on Mars," *Journal of Geophysical Research: Planets* 121, no. 4 (April 1, 2016): 2015JE004972.

43. For contrasting views, consider James L. Fastook and James W. Head, "Glaciation in the Late Noachian Icy Highlands: Ice Accumulation, Distribution, Flow Rates, Basal Melting, and Top-down Melting Rates and Patterns," *Planetary and Space Science* 106 (2015): 82–98; Yo Matsubara, Alan D. Howard, and J. Parker Gochenour, "Hydrology of Early Mars: Valley Network Incision," *Journal of Geophysical Research: Planets* 118, no. 6 (June 1, 2013): 1365–87; and Gaetano Di Achille and Brian M. Hynek, "Ancient Ocean on Mars Supported by Global Distribution of Deltas and Valleys," *Nature Geoscience* 3, no. 7 (July 2010): 459–63.

44. R. E. Milliken et al., "Opaline Silica in Young Deposits on Mars," *Geology* 36, no. 11 (November 1, 2008): 847–50.

45. Diedrich T. F Möhlmann, "Water in the Upper Martian Surface at Mid- and Low-Latitudes: Presence, State, and Consequences," *Icarus* 168, no. 2 (April 2004): 318–23; Lujendra Ojha et al., "Spectral Evidence for Hydrated Salts in Recurring Slope Lineae on Mars," *Nature Geoscience* 8,

no. 11 (November 2015): 829–32; F. Javier Martín-Torres et al., "Transient Liquid Water and Water Activity at Gale Crater on Mars," *Nature Geoscience* 8, no. 5 (May 2015): 357–61.

Significantly, the presence of liquid water, even from seeping, would mean that RSL sites on Mars meet the conditions for being an "Uncertain Region" according to COSPAR's guidelines and should be treated as a Special Region (a region where terrestrial organisms are likely to replicate) going forward. See John D. Rummel et al., "A New Analysis of Mars 'Special Regions': Findings of the Second MEPAG Special Regions Science Analysis Group (SR-SAG2)," *Astrobiology* 14, no. 11 (November 2014): 887–968.

46. See C. T. Adcock, E. M. Hausrath, and P. M. Forster, "Readily Available Phosphate from Minerals in Early Aqueous Environments on Mars," *Nature Geoscience* 6, no. 10 (October 2013): 824–27. They even argue that with regard to a particular problem that could have hindered the formation of life, the prebiotic availability of phosphate, the conditions on Mars may have been *more amenable* than they were on the early Earth.

47. David S. McKay et al., "Search for Past Life on Mars: Possible Relic Biogenic Activity in Martian Meteorite ALH84001," *Science* 273, no. 5277 (August 16, 1996): 924–30.

48. See, for instance, Juan-Manuel Garcia-Ruiz, "Morphological Behavior of Inorganic Precipitation Systems," in *SPIE Proceedings*, vol. 3755, 1999, 74–82; Mikhail Y. Zolotov and E. L. Shock, "An Abiotic Origin for Hydrocarbons in the Allan Hills 84001 Martian Meteorite through Cooling of Magmatic and Impact-Generated Gases," *Meteoritics & Planetary Science* 35, no. 3 (May 2000): 629–38; and K. L. Thomas-Keprta et al., "Origins of Magnetite Nanocrystals in Martian Meteorite ALH84001," *Geochimica et Cosmochimica Acta* 73, no. 21 (November 1, 2009): 6631–77.

49. Recent work suggests that a low-pressure environment may not be as problematic as previously supposed. See R. L. Mickol and T. A. Kral, "Low Pressure Tolerance by Methanogens in an Aqueous Environment: Implications for Subsurface Life on Mars," *Origins of Life and Evolution of Biospheres*, September 23, 2016, 1–22.

50. Lindsey S. Link, Bruce M. Jakosky, and Geoffrey D. Thyne, "Biological Potential of Low-Temperature Aqueous Environments on Mars," *International Journal of Astrobiology* 4, no. 2 (April 2005): 155–64; or Diedrich T. F. Möhlmann, "Are Nanometric Films of Liquid Undercooled Interfacial Water Bio-Relevant?," *Cryobiology* 58, no. 3 (June 2009): 256–61.

## 2. Astrobiology's Intra-Active Aliens

1. Lindsay E. Hays, ed., "Astrobiology Strategy" (NASA, October 2015), https://astrobiology.nasa.gov/uploads/filer_public/01/28/01283266 -e401-4dcb-8e05-3918b21edb79/nasa_astrobiology_strategy_2015_151008 .pdf.

2. Charles Q. Choi, "Mars Life? 20 Years Later, Debate Over Meteorite Continues," Space.com, August 10, 2016, http://www.space.com/33690 -allen-hills-mars-meteorite-alien-life-20-years.html.

3. The satirical website *The Onion* captures this sentiment well. See "Nation Demands NASA Stop Holding Press Conferences Until They Discover Some Little Alien Guys," *The Onion*, September 28, 2015, http:// www.theonion.com/article/nation-demands-nasa-stop-holding-press-conferences-51412.

4. Hays, "Astrobiology Strategy."

5. Lucas John Mix, *Life in Space: Astrobiology for Everyone*, 1st edition (Cambridge: Harvard University Press, 2009), chap. 6. There are alternative, usually energy-first or thermodynamic disequilibrium, approaches that would propose a broader paradigm for life. See Steven A. Benner, Alonso Ricardo, and Matthew A. Carrigan, "Is There a Common Chemical Model for Life in the Universe?," *Current Opinion in Chemical Biology* 8, no. 6 (December 2004): 672–89; and Tori M. Hoehler, Jan P. Amend, and Everett L. Shock, "A 'Follow the Energy' Approach for Astrobiology," *Astrobiology* 7, no. 6 (December 1, 2007): 819–23.

6. Carl Pilcher, "Questing for Life: The Scientific Challenge," Center of Theological Inquiry Winter Symposium 2017, Princeton, N.J., January 30, 2017.

7. NASA has been using variations of this definition for quite some time. The development of this well-known line, usually attributed to Gerald Joyce though stemming from a suggestion of Carl Sagan, is summarized in the foreword to David W. Deamer and Gail R. Fleischacker's work *Origins of Life: The Central Concepts* (Boston: Jones & Bartlett Publishers, 1994). The challenge of defining life for astrobiology continues and two critical contributions to consider are Steven A. Benner, "Defining Life," *Astrobiology* 10, no. 10 (December 2010): 1021–30, and Erik Persson, "Philosophical Aspects of Astrobiology," in *The History and Philosophy of Astrobiology*, ed. David Dunér et al. (Newcastle: Cambridge Scholars Publishing, 2013), 29–48.

8. David Grinspoon, *Earth in Human Hands: Shaping Our Planet's Future* (New York: Grand Central Publishing, 2016), chap. 2.

9. Shawn D. Domagal-Goldman et al., "The Astrobiology Primer v2.0," *Astrobiology* 16, no. 8 (August 1, 2016): 561–653.

10. Dirk Schulze-Makuch and Louis Neal Irwin, *Life in the Universe: Expectations and Constraints* (Berlin: Springer, 2008), chap. 6.

11. See Schulze-Makuch and Irwin, *Life in the Universe*, chaps. 4–5. See also William Bains, "Many Chemistries Could Be Used to Build Living Systems," *Astrobiology* 4, no. 2 (June 2004): 137–67.

12. Tori M. Hoehler, "An Energy Balance Concept for Habitability," *Astrobiology* 7, no. 6 (December 2007): 824–38.

13. C. S. Cockell et al., "Habitability: A Review," *Astrobiology* 16, no. 1 (January 2016): 89–117.

14. See Karen Barad, *Meeting the Universe Halfway: Quantum Physics and the Entanglement of Matter and Meaning* (Durham: Duke University Press Books, 2007), 431n38. I am not convinced the distinction is as dramatic as she indicates.

15. Barad's primary concern is giving an account of philosophical realism that can be made consistent with quantum physics. When she uses the term "phenomena," she has in mind the sense that Niels Bohr intended: Phenomena relate to the laboratory setups used in studying quantum events that create distinctive epistemological parameters inseparable from the hypothesized quantum events. This shift to the *ontological* significance of phenomena is where Barad's agential realism departs most notably from Bohr's work (which remains notably epistemological). Issues of entanglement are the place where theologians (especially) pick up on Barad's work, such that the account of intra-action becomes a means to describing how entanglement gets conceptualized across increasingly complex ontological entities as a metaphor. See, for instance, Catherine Keller, *Cloud of the Impossible: Negative Theology and Planetary Entanglement* (New York: Columbia University Press, 2014), and Catherine Keller and Mary-Jane Rubenstein, eds., *Entangled Worlds: Religion, Science, and New Materialisms* (New York: Fordham University Press, 2017). My aim is something slightly different; I want to leave entanglement to the side and use the concept of intra-action phenomenologically—applying it directly to astrobiological systems as a particular sort of ontologically significant unit.

16. Barad, *Meeting*, 139. The language of observer and observed relates to the quantum reality she has in mind, but the description of the wholeness of phenomena is still relevant to my work.

17. Barad, *Meeting*, 140.

18. Barad, *Meeting*, 3.

19. Ed Turner, "Improbable Life" (Center of Theological Inquiry Colloquiums, Princeton, N.J., May 2, 2017); Eric Smith and Harold J. Morowitz, *The Origin and Nature of Life on Earth: The Emergence of the Fourth Geosphere* (New York: Cambridge University Press, 2016).

20. Nick Stockton, "Elon Musk Announces His Plan to Colonize Mars and Save Us All," *Wired*, September 27, 2016, https://www.wired.com/2016/09/elon-musk-colonize-mars/. In actuality, the prospects for terraforming Mars given present technologies would be a *long* process very unlike fantastical, science fiction imaginings.

21. See "Mars Ice Deposit Holds as Much Water as Lake Superior," NASA/JPL, November 22, 2016, http://www.jpl.nasa.gov/news/news.php?feature=6680.

22. Pilcher, "Questing for Life."

23. See Lisa H. Sideris, "Biosphere, Noosphere, and the Anthropocene: Earth's Perilous Prospects in a Cosmic Context," *Journal for the Study of Religion, Nature and Culture* 11, no. 4 (November 16, 2017): 399–419.

### 3. Being a Living-System

1. Robert Cummings Neville, *Ultimates: Philosophical Theology,* vol. 1 (Albany: SUNY Press, 2014), 15.

2. See Paul Tillich, *Dynamics of Faith* (New York: Harper-Collins, 1957), chap. 3. I have written more extensively about Tillich's notion of symbols in "Comparing Tillich and Rahner on Symbol: Evidencing the Modernist/Postmodernist Boundary," *Bulletin of the North American Paul Tillich Society* 37, no. 2 (Spring 2011): 23–38.

3. Paul Tillich, *Systematic Theology*, 3 vols. (Chicago: University of Chicago Press, 1951–63), 1:211.

4. See Edward Farley, *Deep Symbols: Their Postmodern Effacement and Reclamation* (Valley Forge, Pa.: Trinity Press International, 1996), 3–4, 113–15, and 126n4–6.

5. Neville, *Ultimates*, 67 and 77.

6. Farley, *Deep Symbols*, 4–5 and 21–23.

7. Farley, *Deep Symbols*, 3.

8. Farley, *Deep Symbols*, 6.

9. See Farley, *Deep Symbols*, 7.

10. See Farley, *Deep Symbols*, 8 and 24; and Sallie McFague, *Metaphorical Theology: Models of God in Religious Language* (Philadelphia: Fortress Press, 1982), chap. 1.

11. See Philip Hefner, *The Human Factor* (Minneapolis: Augsburg Fortress Publishers, 2000); Noreen Herzfeld, *In Our Image: Artificial Intelligence and the Human Spirit* (Minneapolis: Fortress Press, 2002); J. Richard Middleton, *The Liberating Image: The Imago Dei in Genesis 1* (Grand Rapids: Brazos Press, 2005); F. LeRon Shults, *Reforming Theological Anthropology: After the Philosophical Turn to Relationality* (Grand Rapids: W. B. Eerdmans, 2003); or Wentzel van Huyssteen, *Alone in the World?: Human Uniqueness in Science and Theology* (Grand Rapids: W. B. Eerdmans, 2006).

12. Consider van Huyssteen, *Alone in the World?*, chaps. 2 and 3.

13. See Olli-Pekka Vainio, "Imago Dei and Human Rationality," *Zygon: Journal of Religion & Science* 49, no. 1 (March 2014): 122–23; Jonathan Jong and Aku Visala, "Three Quests for Human Nature: Some Philosophical Reflections," *Philosophy, Theology, and the Sciences* 1, no. 2 (2014): 146–71.

14. These intra-typological squabbles are often invoked to critique any employment of the *imago Dei* (particularly for ecotheologies) because they

evidence that the doctrine is inextricably tied to a pernicious and damaging anthropocentrism. I have not focused on that line of argument in this work, instead assuming such critiques of anthropocentrism in the *imago Dei* are valid but eventually become muted by interpreting the symbol with astrobiology as a context of engagement.

15. See Alistair McFadyen, "Imaging God: A Theological Answer to the Anthropological Question?," *Zygon: Journal of Religion and Science* 47, no. 4 (December 2012): 918–33; Celia Deane-Drummond, "God's Image and Likeness in Humans and Other Animals: Performative Soul-Making and Graced Nature," *Zygon: Journal of Religion and Science* 47, no. 4 (December 2012): 934–48; Joshua M. Moritz, "Human Uniqueness, the Other Hominids, and 'Anthropocentrism of the Gaps' in the Religion and Science Dialogue," *Zygon: Journal of Religion and Science* 47, no. 1 (March 2012): 65–96; and David Fergusson, "Humans Created According to the Imago Dei: An Alternative Proposal," *Zygon: Journal of Religion and Science* 48, no. 2 (May 2013): 439–53.

16. Paul Tillich, "The Effects of Space Exploration on Man's Condition and Stature," in *The Future of Religions*, ed. Jerald C. Brauer (New York: Harper & Row, 1966), 39–51.

17. This is simply a variation on the idea that there is an asymmetry in interdisciplinary and transdisciplinary inquiries. See Arthur Peacocke, *Theology for a Scientific Age: Being and Becoming—Natural, Divine, and Human* (Minneapolis: Fortress Press, 1993); Robert J. Russell, *Cosmology—from Alpha to Omega: The Creative Mutual Interaction of Theology and Science* (Minneapolis: Fortress Press, 2008); or Agustin Fuentes, "Evolutionary Perspectives and Transdisciplinary Intersections: A Roadmap to Generative Areas of Overlap in Discussing Human Nature," *Theology and Science* 11, no. 2 (May 2013): 106–29.

18. Even a case like that made by Gavin Ortlund for the importance of Gen. 5:1–3 frames the meaning of this text in terms of its ability to serve as an interpretive schema for the earlier reference. See "Image of Adam, Son of God: Genesis 5:3 and Luke 3:38 in Intercanonical Dialogue," *Journal of the Evangelical Theological Society* 57, no. 4 (December 2014): 673–88.

The only notable exception to this ordering is found in the well-reasoned work of Graeme Auld, "Imago Dei in Genesis: Speaking in the Image of God," *The Expository Times* 116, no. 8 (May 2005): 259–62. He contends that references from Chapters 5 and 9 are the precursors and the occurrence in Chapter 1 radicalizes and formalizes implications from these more original texts.

19. Walter Brueggemann, *Genesis*, Interpretation (Atlanta: John Knox Press, 1982), 12. By contrast, one might consider the radical creatureliness of the human being without any sense of exclusion in Psalm 104 or 148. See James Luther Mays, *Psalms*, Interpretation (Louisville: Westminster John Knox Press, 1994), 334 and 445.

20. Brueggemann, *Genesis*, 22–38 and Mays, *Psalms*, 331–37

21. Much ink has been spilled as to whether this divine power to "create" should somehow be restricted to God's action as distinct from the human power to "make," "do," or "cut." Such a distinction is dubious. See Claus Westermann, *Genesis 1–11*, trans. John J. Scullion, Hermeneia (Minneapolis: Fortress Press, 1994), 86 and 98–100. For a defense of the distinctiveness of the verb "to create," see Thomas Finley, "Dimensions of the Hebrew Word for 'Create' (Bara')," *Bibliotheca Sacra* 148, no. 592 (October 1991): 409–23.

22. See Westermann, *Genesis 1–11*, 102–3.

23. See Catherine Keller, *Face of the Deep: A Theology of Becoming* (New York: Routledge, 2002).

24. These terms are directly from Tillich's various descriptions of God. See Tillich, *Systematic Theology*, vol. 1, pt. 2, sec. 2; *Theology of Culture* (New York: Oxford University Press, 1959), 23–30; and *The Courage to Be* (New Haven: Yale University Press, 2000), chap. 6.

25. Brueggemann, *Genesis*, 27.

26. Mays, *Psalms*, 336ff.

27. The cosmogonic descriptions certainly do not articulate an understanding of fittedness in terms of metabolic complexity or organic chemical composition, but to expect as much would be horridly anachronistic.

28. Westermann, 156. He stresses that as part of the cosmogony, the *imago Dei* is reflecting on action taken by God in the creation of a particular type of being. Westermann is echoing Karl Barth, even if Barth's work has been criticized exegetically. See Karl Barth, *Church Dogmatics* (Edinburgh: T&T Clark, 1932–67), vol. 3.1, 85. The potential problem is that on this view, the *imago Dei* risks being instrumentalized as a statement *solely* about the consistency of well-ordered structure evidenced in God's creative action.

29. See also McFadyen, "Imaging God," 918–33.

## 4. The *imago Dei* as a Refractive Symbol

1. H. Wildberger, "ṣelem—image," in *Theological Lexicon of the Old Testament*, ed. Ernst Jenni and Claus Westermann, trans. Mark E. Biddle (Peabody, Ma: Hendrickson Publishers, 1997), 1080–81. Two of the seventeen references in Hebrew come from the Psalms (39:7; 73:20), which seem to reference a different Semitic root; *HALOT* suggests \*ṭlm, cf. Aramaic (further Ugaritic \*ẓlm and Arabic \*ḏlm) meaning "darkness," as opposed to \*ṣlm, cf. Arabic \*ṣlm meaning "to cut, carve," for all other occurrences. Contra *HALOT*, Wildberger proposes that the instances of this word in the Psalms are from the same root and show the "remarkable flexibility that characterizes the term."

2. F. J. Stendebach, "ṣelem," in *Theological Dictionary of the Old Testament [TDOT]*, vol. 2, ed. G. Johannes Botterweck, Heinz-Josef Fabry, and Helmer Ringgren, trans. Douglas W. Stott (Grand Rapids: Eerdmans, 2003), 387–88. Languages that include substantive terms generated from this Proto-Semitic root are Ugaritic, Phoenician, Aramaic (Old Aramaic, Biblical Aramaic, Palmyrene, and Nabatean), Old South Arabian, Middle Hebrew, and Arabic (through an Aramaic loan word).

3. Even our Hebrew lexicons continue this separation of the Genesis passages from other uses of the Hebrew word. In both *HALOT* and *BDB* there is a separate definition of the word as "resemblance" (indicating the connection to *dəmût*), and this definition applies to the Genesis passages only.

4. F. J. Stendebach, "ṣelem," 393. See also Wildberger, "*ṣelem*—image," 1082, and H. D. Preus, "*dāmāh*," in *TDOT*, vol. 3, ed. G. Johannes Botterweck and Helmer Ringgren, trans. Geoffrey W. Bromiley, David E. Green, and John T. Willis (Grand Rapids: Eerdmans, 1978), 259. There has also been significant research on what, if any, significance might be derived from contrasting the translation of the preposition *bet* as *bet normae* (in our image)—which follows closely the chosen translation of *bet* as *kata* in the Septuagint—versus *bet essentiae* (as our image). On this point, see the concise summary by Claudia Welz, "Imago Dei: References to the Invisible," *Studia Theologica* 65 (2011): 74–91; and Angelika Berlejung, *Die Theologie Der Bilder: Herstellung Und Einweihung von Kultbildern in Mesopotamien Und Die Alttestamentliche Bilderpolemik* (Fribourg: Vandenhoeck & Ruprecht, 1998), 308–11.

5. Irenaeus, *Saint Irenaeus of Lyons: Against Heresies*, ed. Alexander Roberts, James Donaldson, and A. Cleveland Coxe (Indiana: Ex Fontibus, 2010), 2.2–3. Clearly, Irenaeus is reading the two accounts of creation within one another.

6. See Irenaeus, *Against Heresies*, 4.4.3. See also 4.37–39.

7. See *Against Heresies*, 2.22.4.

8. As Jarsolav Pelikan aptly summarizes, "Christ became the example for men, as Adam had been the example for Christ; being the Logos of God, Christ was not only the example but the exemplar and prototype of the image of God according to which man had been created." See *The Christian Tradition: A History of the Development of Doctrine*, 5 vols. (Chicago: University of Chicago Press, 1971–1989), 1:145. See also D. Minns, "Irenaeus," *Expository Times* 120, no. 4 (2009): 157–66. Lest the spirit of this way of conceptualizing the *imago Dei* be thought to be a remnant of theologies in bygone eras, one might note the similarity to Irenaeus's position reflected in Pannenberg's proleptic account of the *imago Dei* or approaches that stress the eschatological significance of the *imago Dei* over and against the protological. See Cecilia Echeverría Falla, "La

Imagen de Dios En El Hombre: Consideraciones En Torno a La Cuestión En W. Pannenberg," *Scripta Theologica* 45, no. 3 (December 2013): 737–55; and David Fergusson, "Humans Created according to the "Imago Dei: An Alternative Proposal," *Zygon: Journal of Religion and Science* 48, no. 2 (June 1, 2013): 439–53.

9. In the authentic Pauline letters, the Greek term εἰκών (the same term used to translate ṣelem in the Septuagint) appears five times with a meaning that is related to the *imago Dei*—twice in 1 Corinthians; twice in 2 Corinthians; and once in Romans. (I am not including the use of the term in Romans 1:23, a sixth occurrence, because it deals more specifically with the construction of idols; nor am I including 1 Corinthians 11:7 because of contention concerning whether it is an interpolation.) These references takes two characteristic forms: Christ is the image of God (2 Cor. 4:4), or we are the image of Christ (2 Cor. 3:18, 1 Cor. 15:49, and Rom. 8:29). Making the Christ into the mediating term by which human beings are the image of God has the effect of imbuing a concept derived from creation imagery with eschatological meaning. See Ernst Käsemann, *Perspectives on Paul*, trans. Margaret Kohl (Philadelphia: Fortress Press, 1971), 26ff.

10. The oft cited passage is Irenaus, *Against Heresies*, 5.16.2. However, David Cairns and Wolfhart Pannenberg also point to 5.6.1. See David Cairns, *The Image of God in Man* (London: Collins, 1973), 80; and Wolfhart Pannenberg, *Anthropologie in theologischer Perspektive* (Göttingen: Vandenhoeck & Ruprecht, 1983), 45.

11. Irenaeus, *Against Heresies*, 2.35.4.

12. The introduction of a Christian trichotomy (body, soul, and spirit) in contrast to a pagan dichotomy (body and soul) is one of Irenaeus's critical contributions on this point. See Cairns, *The Image of God in Man*, 84–85; Brunner, *Man in Revolt: A Christian Anthropology*, trans. Olive Wyon (Philadelphia: Westminster Press, 1947), 505; and especially John Lawson, *The Biblical Theology of Saint Irenaeus* (London: Epworth Press, 1948), 206–9.

13. Instead of the exegetical distinction, Augustine offers a set of philosophical distinctions between image, equality, and likeness. See Augustine, *Eighty-three different questions*, trans. David Mosher (Washington, D.C: Catholic University of America Press, 1982), q. 74.

14. Tillich, *Theology of Culture*, 86–88

15. Stendebach, "ṣelem," 391.

16. Wildberger, "ṣelem—image," 1081.

17. Wildberger, "ṣelem—image," 1083. Often scholars deepen this connection with kingship noting that the purpose of our being the image of God is to have "dominion" over the animals and to "subdue" the earth in Genesis 1:26–28 and also in Psalm 8. See, for instance, Westermann, *Genesis 1–11*, 158–60. Perhaps, however, this more specific connection to

kingship is a symptom of the more general participatory quality of *ṣelem* as "symbol."

18. A. R. Millard and P. Bordreuil, "A Statue from Syria with Assyrian and Aramaic Inscriptions," *The Biblical Archeologist* 45 (1982): 135–41. A notable absence from my account of other ancient Near Eastern sources is any consideration of the *Gilgamesh Epic*. As to the value (or lack of value) of this material for considering the image of God, consider Middleton, *The Liberating Image*, 95–99.

19. See Augustine, *Confessions*, trans. Henry Chadwick (Oxford: Oxford University Press, 1998), 13.11.12, and Augustine, *The Trinity*, trans. Edmund Hill, vol. 1:5, *The Works of Saint Augustine* (Brooklyn: New City Press, 1990), 1.1.4.

20. On the chiastic structure of *The Trinity* and the importance of Book VII, see Edmund Hill's introduction to Augustine, *The Trinity*, 25–27.

21. Cairns, *The Image of God in Man*, 99; see also Augustine, *The Trinity*, 15.11.21.

22. I am simply leaving aside the issue of *whether* the idea of God as trinity actually appears in Christian scriptures; for Augustine, such appearance would have been self-evident.

23. His argument concerning the *imago Dei* is set up in parallel to his argument regarding knowledge of other souls: that we know and love the soul of the other on analogy to the presence of the form and truth of the soul (even if the qualities of these souls are different) within ourselves. See *The Trinity*, 8.3.7 and 8.4.9. See also Michael T. McNulty, "Augustine's Argument for the Existence of Other Souls," *Modern Schoolman: A Quarterly Journal of Philosophy* 48 (November 1, 1970): 19–24. For a wider introduction to Augustine's account of souls, see Mary T. Clark, "De Trinitate," in *The Cambridge Companion to Augustine*, ed. Norman Kretzmann and Eleonore Stump (New York: Cambridge University Press, 2001), 97. Ronald Teske, "Augustine's theory of soul," in *The Cambridge Companion to Augustine*, ed. Norman Kretzmann and Eleonore Stump (New York: Cambridge University Press, 2001), 116–23.

24. Augustine, *The Trinity*, 5.1.5–6. See also 15.1.5.

25. Augustine, *The Trinity*, 5.2.9. See also Sarah Heaner Lancaster, "Divine Relations of the Trinity: Augustine's Answer to Arianism," *Calvin Theological Journal* 34, no. 2 (November 1, 1999): 327–46.

26. Augustine, *The Trinity*, 8.1.2–3.

27. The impetus to search out this internal Trinitarian image is also scripturally rooted. See Augustine, *On Genesis*, trans. Edmund Hill, vol. 1:13, The Works of Saint Augustine (Hyde Park: New City Press, 2002), 3.19–21.

28. Augustine, *The Trinity*, 8.5.10.

29. Augustine, *The Trinity*, 8.5.12.

30. Augustine, *The Trinity*, 8.5.14.

31. Augustine goes much further than I have here in order to avoid the trap in love by which the lover, the object of love, and the love itself no longer form a trinity but a binary through an act of self-love. He is dedicated to finding a form of relative predication that will always give itself over in threes. See especially, Augustine, *The Trinity*, 9.2.9–9.3.18; 10.3.11–13; and 14.2.1–14.5.24.

32. Claude Welch, *Protestant Thought in the Nineteenth Century*, 2 vols. (Eugene: Wipf and Stock, 1972–85), vol. 1, chap. 2.

33. Friedrich Schleiermacher, *The Christian Faith*, trans. H. R. Mackintosh and J. S. Stewart (Edinburgh: T&T Clark, 1960), §28.2.

34. Schleiermacher, *The Christian Faith*, §3.2–5.

35. See also Richard R. Niebuhr, *Schleiermacher on Christ and Religion* (London: SCM Press, 1965), 121ff.

36. Schleiermacher, *The Christian Faith*, §15. See also the immensely helpful diagram in Welch, *Protestant Thought in the Nineteenth Century*, vol. 1, 74–75.

37. Schleiermacher, *The Christian Faith*, §30. See also Schleiermacher's concern for the centrality of Christ and the work of redemption to Christian systematic theology in §11.

38. Schleiermacher, *The Christian Faith*, §57.1.

39. Schleiermacher, *The Christian Faith*, §4.1.

40. Schleiermacher, *The Christian Faith*, §4.3.

41. Schleiermacher, *The Christian Faith*, §4.4 and 5, Postscript.

42. Schleiermacher, *The Christian Faith*, §60.1.

43. Still, if there is greater dogmatic weight to be given to one over the other, it is to the perfection of humankind because the perfection of the world only has meaning in relation to human being, but the perfection of humankind has its meaning in relation to God as the presence of God-consciousness in humankind. Schleiermacher, *The Christian Faith*, §58.1–2.

44. Schleiermacher, *The Christian Faith*, §61.3–5.

45. I want to highlight that the picture I am drawing of the divine here is in some ways quite minimal. A biblical scholar or confessional theologian could certainly, and perhaps rightly, claim we can say more about what or who we understand the divine we image to be. One can clearly make the case that we need not limit the concept of the divine to what is revealed of God as a cosmogonic creative force or that there is more to say about this God of the cosmogonies than what I am offering here. But this is where interpreting this symbol, *imago Dei*, with astrobiology as the context of engagement becomes important.

46. Technically, when we see our reflection in a mirror, there are two sorts of reflections at work. Diffuse reflection, incident light waves bouncing off irregular surfaces—bodies, clothing, and so on—at many angles, allows us to see. Subsequently, some of the diffuse reflected waves bouncing

off us will encounter the mirror; the silver lining of the mirror allows for specular reflection. In both cases, we are dealing with the rebounding of waves off a surface.

47. Philipp Melanchthon, *A Melanchthon Reader*, trans. Ralph Keen (New York: P. Lang, 1988), 283.

48. I have engaged this issue more directly in Adam Pryor, "Intelligence, Non-Intelligence . . . Let's Call the Whole Thing Off," *Theology and Science* 16, no. 4 (October 2, 2018): 471–83.

## 5. Conceptualizing Nature

1. Though earlier usage is noted, the term is usually traced to Paul J. Crutzen, "Geology of Mankind," *Nature* 415 (January 3, 2002): 23; see also Will Steffen, Paul J. Crutzen, and John R. McNeill, "The Anthropocene: Are Humans Now Overwhelming the Great Forces of Nature?," *AMBIO: A Journal of the Human Environment* 36, no. 8 (December 1, 2007): 614–21. Various starting dates for the Anthropocene have subsequently been suggested. Each has great effect on how the term is conceptualized: for connecting the Anthropocene to the human domestication of animals, see Bruce D. Smith and Melinda A. Zeder, "The Onset of the Anthropocene," *Anthropocene* 4 (December 1, 2013): 8–13; for marking its beginning with the rise of the industrial revolution, a shift at the "Great Acceleration" in the 1950s, and a third stage beginning with global awareness around 2006, see Will Steffen et al., "The Anthropocene: Conceptual and Historical Perspectives," *Philosophical Transactions of the Royal Society of London A: Mathematical, Physical and Engineering Sciences* 369, no. 1938 (March 13, 2011): 842–67.

In preferring the stratigraphic language, I am most closely following the proposal that builds on Crutzen's 2002 article and led to the founding of the Working Group on the Anthropocene described in J. Zalasiewicz et al., "Are We Now Living in the Anthropocene?," *GSA Today* 18, no. 2 (February 2008): 4–8. This puts the Anthropocene beginning around the same time as the "Great Acceleration." See also Will Steffen et al., "The Trajectory of the Anthropocene: The Great Acceleration," *The Anthropocene Review* 2, no. 1 (April 1, 2015): 81–98. However, this is not to discount the importance of the wider history of anthropogenic change to the environment or, particularly, the attitudes and practices of modernity. See especially Christophe Bonneuil and Jean-Baptiste Fressoz, *The Shock of the Anthropocene: The Earth, History and Us* (London: Verso, 2016), 53–54.

2. Damian Carrington, "The Anthropocene Epoch: Scientists Declare Dawn of Human-Influenced Age," *The Guardian*, August 29, 2016, sec. Science. Jeremy Davies provides a brief but helpful overview in his *The Birth of the Anthropocene* (Oakland: University of California Press, 2016), 46–56. For a more extensive treatment, see Christophe Bonneuil, "The Geological Turn: Narratives of the Anthropocene," in *The Anthropocene*

*and the Global Environmental Crisis: Rethinking Modernity in a New Epoch*, ed. Clive Hamilton, Christophe Bonneuil, and Francois Gemenne (Abingdon, UK: Routledge, 2015), 17–31.

3. Jedediah Purdy, *After Nature: A Politics for the Anthropocene* (Cambridge: Harvard University Press, 2015), 2–3.

4. Purdy, *After Nature*, 22.

5. Forrest Clingerman, "Place and the Hermeneutics of the Anthropocene," *Worldviews: Global Religions, Culture, and Ecology* 20, no. 3 (January 1, 2016): 228–29. If there is a difference between Clingerman's hermeneutical concept and my approach to the Anthropocene as a symbol, it is minor indeed. He locates the importance of the hermeneutical concept mediating between the material and ontological levels of change at which meaning emerges (231). My own use of symbol and intra-action is meant to convey that the material and the ontological are *so* interconnected that their being parsed apart is important but secondary. The primordial ontological unit is human being and the world in intra-action or as mutually constituted in givenness. As such, I am sympathetic to the aim of Christina Gschwandtner in her exploration of nature as a saturated phenomenon that in its givenness is a counter-experience to the self. See Christina M. Gschwandtner, "Might Nature Be Interpreted as a 'Saturated Phenomenon,'?" in *Interpreting Nature: The Emerging Field of Environmental Hermeneutics*, ed. Forrest Clingerman et al. (New York: Fordham University Press, 2013), 82–101.

6. Richard Monastersky, "Anthropocene: The Human Age," *Nature* 519, no. 7542 (March 12, 2015): 144–47.

7. Paul J. Crutzen and Christian Schwägerl, "Living in the Anthropocene: Toward a New Global Ethos," *Yale Environment 360*, January 24, 2011.

8. Dipesh Chakrabarty, "The Climate of History: Four Theses," *Critical Inquiry* 35, no. 2 (2009): 197–222.

9. Bonneuil and Jean-Baptiste Fressoz, *The Shock of the Anthropocene*, 39.

10. Paul Tillich, *Systematic Theology*, 3 vols. (Chicago: University of Chicago Press, 1951–63), 1:211.

11. On this idea of technical reason, I have in mind Tillich's distinction between ontological and technical reason or Erazim Kohák's observation that nature helps us see past the dualism of *techne* and *poiesis*. See Tillich, *Systematic Theology*, 1:71–75; and Erazim V. Kohák, *The Embers and the Stars: A Philosophical Inquiry into the Moral Sense of Nature* (Chicago: University of Chicago Press, 1984), 17 and 32.

12. Sara Ahmed, *Queer Phenomenology: Orientations, Objects, Others* (Durham, N.C.: Duke University Press, 2006), 166.

13. There are any number of existential theologies that address anxiety in this way following Søren Kierkegaard in *The Concept of Anxiety: A*

*Simple Psychologically Orienting Deliberation on the Dogmatic Issue of Hereditary Sin*, trans. Reidar Thomte and Albert B. Anderson (Princeton: Princeton University Press, 1981). Usually they clarify how anxiety leads to sin. I will put this aspect of anxiety aside in order to focus on how disorientation might generate anxiety. In short, instead of focusing on anxiety as a theological precursor to sin, I will focus on disorientation as a precursor to anxiety.

14. Amitav Ghosh, *The Great Derangement* (Chicago: University of Chicago Press, 2016), 33, 54, 66, and 111. The derangement subsequent to disorientation in Ghosh's work refers to a bankruptcy in our ability to engage imagination.

15. Ghosh, *The Great Derangement*, 60, 65, and 129–35.

16. Davies, *The Birth of the Anthropocene*, 21. Quoting Chris Caseldine, "Conceptions of Time in (Paleo)climate Science and Some Implications," *Wiley Interdisciplinary Reviews: Climate Change* 3, no. 4 (2012): 334.

17. Davies, *The Birth of the Anthropocene*, 23.

18. Here I take Purdy's analysis of the American imaginary to be important because of the ways it has informed popular sensibilities about nature.

19. C. S. Lewis, *Studies in Words* (Cambridge: Cambridge University Press, 1960), 24–74. Of course, other notable works have covered this territory as well, and in more depth, as a critique of nature understood as that which is without human interference. See Bruno Latour, *Politics of Nature: How to Bring the Sciences into Democracy*, trans. Catherine Porter (Cambridge: Harvard University Press, 2004); Jane Bennett, *Vibrant Matter: A Political Ecology of Things* (Durham, N.C.: Duke University Press, 2010); and Donna J. Haraway, *Staying with the Trouble: Making Kin in the Chthulucene* (Durham, N.C.: Duke University Press, 2016).

20. Lewis contends it is only a short jump from such categorizations of "natures" to an understanding of "Nature" designating the whole totality of existence, or what he calls "*nature* in the dangerous sense." He suggests that it is only rarely that Nature is actually used this way; far more often nature is still designating some class or quality of things as a contrast. See *Studies in Words*, 37–40.

21. Lewis, *Studies in Words*, 46.

22. Purdy, *After Nature*, 17–20.

23. Purdy, *After Nature*, 21, emphasis mine. Purdy's point easily translates into a more explicitly theological formulation. See Peter Scott, *A Political Theology of Nature* (Cambridge and New York: Cambridge University Press, 2003), pt. 1.

24. Consider also Donna Haraway, "A Cyborg Manifesto: Science, Technology, and Socialist-Feminism in the Late Twentieth Century," in *Simians, Cyborgs, and Women: The Reinvention of Nature* (New York:

Routledge, 1991), 149–82; S. Franklin, "Science as Culture, Cultures as Science," *Annual Review of Anthropology* 24 (1995): 163–84; Anne Kull, "Speaking Cyborg: Technoculture and Technonature," *Zygon* 37, no. 2 (June 2002): 279–87; and Noreen Herzfeld, *Technology and Religion: Remaining Human in a Co-Created World* (Philadelphia: Templeton Press, 2009).

25. Bonneuil and Fressoz, *The Shock of the Anthropocene*, chaps. 3 and 4.

26. Specifically, the connection would be that the story of American-style democracy is crucial to understanding environmental reform as an intensified politics of nature in a post-humanist democracy of self-restraint. See Purdy, *After Nature*, 266–72.

By no means does Purdy's analysis lead inevitably to the specifics of his conclusion. One might equally well make the case that his historical analysis could lead to a politics of "Social Flesh" as with Chris Beasley and Carol Bacchi, "Envisaging a New Politics for an Ethical Future: Beyond Trust, Care and Generosity—Towards an Ethic of "Social Flesh," *Feminist Theory* 8, no. 3 (December 1, 2007): 279–98; and Sharon V. Betcher, *Spirit and the Obligation of Social Flesh: A Secular Theology for the Global City* (New York: Fordham University Press, 2013).

27. Purdy, *After Nature*, 107–11.

28. Purdy, *After Nature*, 82–83. This is a version of the "agricultural argument." Thomas Flanagan, importantly, contends that use of the argument may not have been as prevalent as previously supposed. See Thomas Flanagan, "The Agricultural Argument and Original Appropriation: Indian Lands and Political Philosophy," *Canadian Journal of Political Science* 22, no. 3 (September 1989): 589–602.

On the other hand, the categories of harmony and beauty are indicative of a long tradition by which we understand authority that is nearly ubiquitous in modern times. Consider Mary Ellen O'Connell, "Law, Theology, and Aesthetics: Identifying the Sources of Authority," in *Theology as Interdisciplinary Inquiry: Learning with and from the Natural and Human Sciences*, ed. Robin W. Lovin and Joshua Mauldin (Grand Rapids: W. B. Eerdmans, 2017), 112–31.

29. The development of an implicit harmony of moral order through human cultivation is well attested, perhaps most notably in Georges Louis Leclerc Buffon, *Les époques de la nature* (Paris: De l'imprimerie royale, 1780).

30. Purdy, *After Nature*, 134.

31. Purdy, *After Nature*, 141–44.

32. In a sense, Marsh's conservationist approach opens the door to contemporary environmental ethics in important and lasting ways. See David Lowenthal, *George Perkins Marsh: Prophet of Conservation* (Seattle: University of Washington Press, 2009); and David Lowenthal, "Nature

and Morality from George Perkins Marsh to the Millennium," *Journal of Historical Geography* 26, no. 1 (January 1, 2000): 3–23.

33. Purdy, *After Nature*, 163.

34. Purdy, *After Nature*, 201.

35. Purdy, *After Nature*, 189–91.

36. Purdy, *After Nature*, 216–17.

37. Purdy, *After Nature*, 224.

38. On the problems with environmental cost-benefit analysis, see Frank Ackerman and Lisa Heinzerling, "Pricing the Priceless: Cost-Benefit Analysis of Environmental Protection," *University of Pennsylvania Law Review* 150, no. 5 (2002): 1553–84; Frank Ackerman and Lisa Heinzerling, *Priceless: On Knowing the Price of Everything and the Value of Nothing* (New York: The New Press, 2005); and Maria Damon, Kristina Mohlin, and Thomas Sterner, "Putting a Price on the Future of Our Children and Grandchildren," in *The Globalization of Cost-Benefit Analysis in Environmental Policy*, ed. Michael A. Livermore and Richard L. Revesz (Oxford: Oxford University Press, 2013), chapter 4.

39. Purdy, *After Nature*, 238–39.

40. Bonneuil and Fressoz, *The Shock of the Anthropocene*, 170–72.

41. Bonneuil and Fressoz, *The Shock of the Anthropocene*, 196–97.

42. If we do not intentionally take an active responsibility for nature then we will likely return to a default tendency to ignore our environmental reflexivity. Purdy, *After Nature*, 284–86.

43. Ben Dibley, "'Nature Is Us': The Anthropocene and Species-Being," *Transformations* 21 (2012): special section, pt. 1; Eileen Crist, "On the Poverty of Our Nomenclature," *Environmental Humanities* 3 (2013): 129–47; Andreas Malm and Alf Hornborg, "The Geology of Mankind? A Critique of the Anthropocene Narrative," *Anthropocene Review* 1, no. 1 (2014): 62–69; Jeremy Baskin, "Paradigm Dressed as Epoch: The Ideology of the Anthropocene," *Environmental Values* 24, no. 1 (2015): 9–29; Donna J. Haraway, *Staying with the Trouble*.

44. Purdy, *After Nature*, 229.

45. Davies, *The Birth of the Anthropocene*, 62.

### 6. The Anthropocene as Planetarity in Deep Time

1. Jedediah Purdy, *After Nature: A Politics for the Anthropocene* (Cambridge: Harvard University Press, 2015), chaps. 7–8.

2. Andrea Wulf, *The Invention of Nature: Alexander von Humboldt's New World* (New York: Knopf, 2015); Robert J. Richards, *The Romantic Conception of Life: Science and Philosophy in the Age of Goethe* (Chicago: University of Chicago Press, 2004); Kate Rigby, *Topographies of the Sacred: The Poetics of Place in European Romanticism* (Charlottesville: University of Virginia Press, 2004). Perhaps the best example in this regard

is Wulf's work. In her account, Alexander von Humboldt's sketch of the *Naturgemälde* that eventually appeared in his *Essay on the Geography of Plants* becomes a token for the subsequent ecological development of a "web of life" that prefigures the intra-active sensibility now prevalent in the Anthropocene.

3. Bruno Latour, *Politics of Nature: How to Bring the Sciences into Democracy*, trans. Catherine Porter (Cambridge: Harvard University Press, 2004); Timothy Morton, *Ecology without Nature: Rethinking Environmental Aesthetics* (Cambridge: Harvard University Press, 2009); Jane Bennett, *Vibrant Matter: A Political Ecology of Things* (Durham, N.C.: Duke University Press, 2010); Slavoj Žižek, *In Defense of Lost Causes* (London: Verso, 2017).

4. Gayatri Chakravorty Spivak, *Death of a Discipline* (New York: Columbia University Press, 2003), chap. 3. Initially, the idea appears in a recently republished lecture; see "Imperative to Re-Imagine the Planet," in *An Aesthetic Education in the Era of Globalization* (Cambridge: Harvard University Press, 2013), 335–50.

5. Gayatri Chakravorty Spivak, "'Planetarity' (Box 4, *WELT*)," *Paragraph* 38, no. 2 (July 2015): 290. See also Stephen D. Moore, "Situating Spivak," in *Planetary Loves: Spivak, Postcoloniality, and Theology*, ed. Stephen D. Moore and Mayra Rivera (New York: Fordham University Press, 2010), 27.

6. Spivak, "Imperative to Re-Imagine the Planet," 338.

7. Whitney A. Bauman, *Religion and Ecology: Developing a Planetary Ethic* (New York: Columbia University Press, 2014), 3. An echo of the providential approach to nature is clearly present in global thinking.

8. See Spivak, "'Planetarity' (Box 4, *WELT*)," 291; *Death of a Discipline*, 73; and "Imperative to Re-Imagine the Planet," 338–39 and 350.

9. Spivak, "'Planetarity' (Box 4, *WELT*)," 291. Or, more poetically, "alterity remains underived from us; it is not our dialectical negation, it contains us as much as it flings us away." *Death of a Discipline*, 73.

10. Here I take "self-evident" in the sense described by Edmund Husserl. See *Formal und transzendentale Logik. Versuch einer Kritik der logischen Vernunft. Mit ergäzenden Texten*, ed. Paul Janssen, vol. 17, Husserliana (The Hague: Martinus Nijhoff, 1974), §104–7.

11. Bauman, *Religion and Ecology*, 56–61.

12. See Spivak, "'Planetarity' (Box 4, *WELT*)," 292; and Don McKay, "Ediacaran and Anthropocene: Poetry as a Reader for Deep Time," in *Making the Geologic Now*, ed. Elizabeth Ellsworth and Jamie Kruse (New York: Punctum, 2013), 46–54.

13. Donna Haraway, "Situated Knowledges," in *Simians, Cyborgs, and Women: The Reinvention of Nature* (New York: Routledge, 1991), 183–202.

14. Kwok Pui-Lan, "What Has Love to Do with It? Planetarity, Feminism, and Theology," in *Planetary Loves: Spivak, Postcoloniality, and Theology*, ed. Stephen D. Moore and Mayra Rivera (New York: Fordham University Press, 2010), 38.

15. Spivak, "Imperative to Re-Imagine the Planet," 349. In the same essay (p. 341), she describes this in terms of the Islamic concept of *Haq*.

16. Bauman is clear we must destabilize accounts of religion as well as nature. In destabilizing foundationalist understandings of religion, Bauman contends religions are biohistorical meaning-making practices arising from particular natural-cultural contexts that they in turn shape as a "third-order" theology (e.g., the task of theology is to *deliberately* construct accounts of the divine that provide a symbolic framework for the meaning of human existence). See Bauman, *Religion and Ecology*, 71–76. On "third-order theology," see Gordon D. Kaufman, *An Essay on Theological Method*, 3rd ed. (Atlanta: Oxford University Press, 1995).

17. Bauman, *Religion and Ecology*, 25 and 41.

18. Bauman, *Religion and Ecology*, 25, 38, and 140.

19. Theologically this tends toward panentheism. The immanence of nature gives rise to a "third-order" theology that deliberately constructs an account of divine transcendence in non-separable dialectical tension with nature itself.

20. Bauman, *Religion and Ecology*, 105. See also Catherine Keller, *Cloud of the Impossible: Negative Theology and Planetary Entanglement* (New York: Columbia University Press, 2014), chap. 5. In the interest of clarity, I will not make direct reference to Gilles Deleuze's work to avoid introducing yet another dense set of philosophical terms. Implicitly, I am making extensive use of *Difference and Repetition*, trans. Paul Patton (New York: Columbia University Press, 1995).

21. Keller, *Cloud of the Impossible*, 177 and 191; Clayton Crockett, *Deleuze Beyond Badiou: Ontology, Multiplicity, and Event* (New York: Columbia University Press, 2013), 33–39.

22. There is a resonance between the non-separable difference of "folding" and the indeterminate relationality of the "intra-action" of a "phenomenon" in Karen Barad's agential realism. The way that each term is being used here, it does seem they could be fruitfully woven together—a process Bauman, in terms of performative agency, and Keller, in terms of entanglement, have each begun. See Keller, *Cloud of the Impossible*, chap. 4; and Bauman, *Religion and Ecology*, chaps. 4 and 5.

23. See Robert Mugerauer, "Layering: Body, Building, Biography," in *Interpreting Nature: The Emerging Field of Environmental Hermeneutics*, ed. Forrest Clingerman et al. (New York: Fordham University Press, 2013), 65–69.

24. Wesley J. Wildman, "Distributed Identity: Human Beings as Walking, Thinking, Ecologies in the Microbial World," in *Human Identity at*

*the Intersection of Science, Technology, and Religion*, ed. Nancey Murphy and Christopher C. Knight (Burlington: Ashgate, 2010).

25. See also "dermal metaphysics" in Adam Pryor, *Body of Christ Incarnate for You: Conceptualizing God's Desire for the Flesh* (Lanham, Md.: Lexington, 2016), chaps. 5 and 8.

26. Bauman, *Religion and Ecology*, 116.

27. Bauman, *Religion and Ecology*, 121–23.

28. Bauman, *Religion and Ecology*, 125.

29. Bauman develops this idea with his "polyamory of place" or an "ethics of movement." The key to his ethics of movement is to render our ethical decision in terms of a more complex account of agency and causality. See *Religion and Ecology*, 133ff.

30. See Dylan Trigg, "Bodily Moods and Unhomely Environments: The Hermeneutics of Agoraphobia and the Spirit of Place," in *Interpreting Nature: The Emerging Field of Environmental Hermeneutics*, ed. Forrest Clingerman et al. (New York: Fordham University Press, 2013), 160–77.

31. Bauman, *Religion and Ecology*, 163–64.

32. Jeremy Davies, *The Birth of the Anthropocene* (Oakland: University of California Press, 2016), 23.

33. Christophe Bonneuil and Jean-Baptiste Fressoz, *The Shock of the Anthropocene: The Earth, History and Us* (London: Verso, 2016), 37. Summarizing the methodology found in Fernand Braudel, *The Mediterranean and the Mediterranean World in the Age of Philip II*, 2 vols. (London: Fontana, 1966).

34. Aldo Leopold, *A Sand County Almanac, and Sketches Here and There* (Oxford: Oxford University Press, 1989).

35. See also Adam Frank and Woodruff Sullivan, "Sustainability and the Astrobiological Perspective: Framing Human Futures in a Planetary Context," *Anthropocene* 5 (March 1, 2014): 32–41; "A New Empirical Constraint on the Prevalence of Technological Species in the Universe," *Astrobiology* 16, no. 5 (April 22, 2016): 359–62; and Adam Frank, *Light of the Stars: Alien Worlds and the Fate of the Earth* (New York: Norton, 2018).

36. David Grinspoon, *Earth in Human Hands: Shaping Our Planet's Future* (New York: Grand Central Publishing, 2016), xv.

37. Bonneuil and Fressoz, *Shock of the Anthropocene*, 86.

38. See Davies, *The Birth of the Anthropocene*, 108, 114, 126, and 145.

39. Kees Boeke, *Cosmic View: The Universe in 40 Jumps* (New York: John Day Company, 1957); Charles Eames and Ray Eames, *The Powers of Ten*, Short Documentary (IBM, 1977).

40. Grinspoon, *Earth in Human Hands*, xiv.

41. Grinspoon, *Earth in Human Hands*, 263–65. There are less extreme changes in planetary history; the Anthropocene could be a new "period" or a new "era."

42. Grinspoon, *Earth in Human Hands*, 271.

43. Grinspoon, *Earth in Human Hands*, 272–73.

44. Grinspoon, *Earth in Human Hands*, 277.

45. Pierre Teilhard de Chardin, *The Phenomenon of Man*, trans. Bernard Wall (New York: Harper Perennial, 1976).

46. Anthony D. Barnosky et al., "Has the Earth's Sixth Mass Extinction Already Arrived?," *Nature* 471, no. 7336 (March 3, 2011): 51–57; Elizabeth Kolbert, *The Sixth Extinction: An Unnatural History* (New York: Picador, 2015).

47. Grinspoon, *Earth in Human Hands*, 320–21. This risk that homo sapiens may not induce a Sapiezoic eon, is crucial to why Grinspoon's proposal does not fall prey to the excellent critique offered by Lisa Sideris of planetary usury and interstellar escapism encouraged by many brands of technophilic cosmism in "Biosphere, Noosphere, and the Anthropocene: Earth's Perilous Prospects in a Cosmic Context," *Journal for the Study of Religion, Nature and Culture* 11, no. 4 (November 16, 2017): 399–419.

48. The discovery of exoplanets and other solar systems has the potential to enrich drastically the number of bodies that might be part of this interplanetary comparison regarding the function and development of climate as it relates to habitability. See Grinspoon, *Earth in Human Hands*, 50–54. Perhaps one of our first points of comparison will be Gliese 1132 b if the recently discovered likelihood of an atmosphere on this exoplanet withstands further scientific scrutiny. See John Southworth et al., "Detection of the Atmosphere of the 1.6 $M_\oplus$ Exoplanet GJ 1132 b," *The Astronomical Journal* 153, no. 4 (2017): 191.

49. Grinspoon, *Earth in Human Hands*, 88. This is squarely in line with proposals in geological history emphasizing "neocatastrophism."

50. Grinspoon, *Earth in Human Hands*, 114.

51. Carl Sagan and George Mullen, "Earth and Mars: Evolution of Atmospheres and Surface Temperatures," *Science* 177, no. 4043 (July 7, 1972): 52–56; James C. G. Walker, P. B. Hays, and J. F. Kasting, "A Negative Feedback Mechanism for the Long-Term Stabilization of Earth's Surface Temperature," *Journal of Geophysical Research: Oceans* 86, no. C10 (October 20, 1981): 9776–82; J. William Schopf, *Earth's Earliest Biosphere: Its Origin and Evolution* (Princeton: Princeton University Press, 1983); Alexander A. Pavlov et al., "Greenhouse Warming by CH4 in the Atmosphere of Early Earth," *Journal of Geophysical Research: Planets* 105, no. E5 (May 25, 2000): 11981–90; A. P. Numan et al., "Waves and Weathering on the Early Earth: Geological Evidence for an Equable Terrestrial Climate at 3.7 GA" (Third International Conference on Early Mars, Lake Tahoe, Nevada, 2012), http://www.lpi.usra.edu/meetings/earlymars2012/pdf/7062.pdf. However, recent findings question how much temperature change slows down weathering reactions. See Joshua Krissansen-Totton and David C. Catling, "Constraining Climate Sensitivity and Continental

versus Seafloor Weathering Using an Inverse Geological Carbon Cycle Model," *Nature Communications* 8 (May 22, 2017): 15423.

52. Grinspoon, *Earth in Human Hands*, 41–50.

53. See Steven Earle, *Physical Geology*, 1.5 (BC Open Textbook Project, 2015), https://opentextbc.ca/geology/, sec. 5.6.

54. James F. Kasting, "Runaway and Moist Greenhouse Atmospheres and the Evolution of Earth and Venus," *Icarus* 74, no. 3 (June 1, 1988): 472–94; David Grinspoon, *Venus Revealed: A New Look Below The Clouds Of Our Mysterious Twin Planet* (Cambridge, Mass.: Perseus Publishing, 1997); F. Nimmo and D. McKenzie, "Volcanism and Tectonics on Venus," *Annual Review of Earth and Planetary Sciences* 26, no. 1 (May 1, 1998): 23–51; M. J. Way et al., "Was Venus the First Habitable World of Our Solar System?," *Geophysical Research Letters* 43, no. 16 (August 28, 2016): 8376–83.

55. Bruce M. Jakosky et al., "Initial Results from the MAVEN Mission to Mars," *Geophysical Research Letters* 42, no. 21 (November 16, 2015): 2015GL065271; B. M. Jakosky et al., "MAVEN Observations of the Response of Mars to an Interplanetary Coronal Mass Ejection," *Science* 350, no. 6261 (November 6, 2015): aad0210.

56. If the recent findings of organic compounds on Mars contribute to evidence that a second biogenesis event occurred on that planet, then this comparison becomes more fascinating. Jennifer L. Eigenbrode et al., "Organic Matter Preserved in 3-Billion-Year-Old Mudstones at Gale Crater, Mars," *Science* 360, no. 6393 (June 8, 2018): 1096–1101.

57. Davies, *The Birth of the Anthropocene*, 60.

58. Grinspoon, *Earth in Human Hands*, 77. Of course this echoes the Gaia hypothesis of James Lovelock and Lynn Margulis. J. E. Lovelock, "Gaia as Seen through the Atmosphere," *Atmospheric Environment (1967)* 6, no. 8 (August 1, 1972): 579–80; James E. Lovelock and Lynn Margulis, "Atmospheric Homeostasis by and for the Biosphere: The Gaia Hypothesis," *Tellus* 26, no. 1–2 (January 1, 1974): 2–10

59. See also Eric Smith and Harold J. Morowitz, *The Origin and Nature of Life on Earth: The Emergence of the Fourth Geosphere* (New York: Cambridge University Press, 2016).

60. David Catling, *Astrobiology: A Very Short Introduction* (Oxford: Oxford University Press, 2013), chap. 4; Martha E. Sosa Torres, Juan P. Saucedo-Vázquez, and Peter M. H. Kroneck, "The Magic of Dioxygen," in *Sustaining Life on Planet Earth: Metalloenzymes Mastering Dioxygen and Other Chewy Gases*, Metal Ions in Life Sciences (Springer, 2015), 1–12.

61. Davies, *The Birth of the Anthropocene*, chap. 5.

62. The Earth has rebounded from mass extinctions and will likely continue to be a living planet, evolving new forms of complex life. The longer-term implications of the Anthropocene matter most to *us*: those species living on the planet today who will be subject to extinction.

63. Grinspoon, *Earth in Human Hands*, 115.

64. Grinspoon, *Earth in Human Hands*, 142.

65. Consider also Bronislaw Szerszynski, "Viewing the Technosphere in an Interplanetary Light," *The Anthropocene Review*, October 19, 2016; and "Planetary Mobilities: Movement, Memory and Emergence in the Body of the Earth," *Mobilities* 11, no. 4 (August 7, 2016): 614–28.

66. See also Peter Haff, "Humans and Technology in the Anthropocene: Six Rules," *The Anthropocene Review* 1, no. 2 (August 1, 2014): 126–36; P. K. Haff, "Technology as a Geological Phenomenon: Implications for Human Well-Being," *Geological Society of London Special Publications* 395 (May 1, 2014): 301–9.

67. Davies, *The Birth of the Anthropocene*, 207.

68. Bonneuil and Fressoz, *The Shock of the Anthropocene*, chap. 10.

69. The only difference between climate change due to increased carbon dioxide and the great oxygenation event may be that in one case the byproduct producing climate change comes directly from the bodies of various organisms (the atmospheric oxygen produced through cyanobacteria by photosynthesis) versus being produced to better meet the needs of organisms (the carbon dioxide produced by human beings through industrial production). See Davies, *The Birth of the Anthropocene*, 206.

70. Grinspoon, *Earth in Human Hands*, 226. The previous analysis of nature and the history of environmental reflexivity can nuance Grinspoon's development of the "proto-Anthropocene." Nonetheless, the term is helpful in distinguishing what he sees to be at stake by situating the Anthropocene into deep time.

71. Grinspoon, *Earth in Human Hands*, 225.

72. Grinspoon, *Earth in Human Hands*, 474–49.

73. Grinspoon, *Earth in Human Hands*, 422.

74. Grinspoon, *Earth in Human Hands*, 412.

## 7. An Artful Planet

1. See J. Wentzel van Huyssteen in "What Makes Us Human? The Interdisciplinary Challenge to Theological Anthropology and Christology," *Toronto Journal of Theology* 26, no. 2 (September 1, 2010): 143–44; and Delwin Brown, "Public Theology, Academic Theology: Wentzel Van Huyssteen and the Nature of Theological Rationality," *American Journal of Theology & Philosophy* 22, no. 1 (2001): 88–101.

2. Though not framed in terms of the *imago Dei*, something similar is stressed by Adam Frank. See his "Climate Change and the Astrobiology of the Anthropocene," *Cosmos & Culture: Commentary on Science and Society* 13, no. 7 (October 1, 2016), http://www.npr.org/sections/13.7/2016/10/01/495437158/climate-change-and-the-astrobiology-of-the-anthropocene; and *Light of the Stars: Alien Worlds and the Fate of the Earth* (New York: Norton, 2018).

3. Erazim V. Kohák, *The Embers and the Stars: A Philosophical Inquiry into the Moral Sense of Nature* (Chicago: University of Chicago Press, 1984), 39–42 and 54–55.

4. "Where Does It Hurt?," *On Being* (American Public Media, September 15, 2016), https://onbeing.org/programs/ruby-sales-where-does-it-hurt/.

5. Christophe Bonneuil and Jean-Baptiste Fressoz, *The Shock of the Anthropocene: The Earth, History and Us* (London: Verso, 2016), 107–20 and 157–61.

### 8. Living-Into Presence, Wonder, and Play

1. See Adam Pryor, *The God Who Lives: Investigating the Emergence of Life and the Doctrine of God* (Eugene: Pickwick Publications, 2014); and *Body of Christ Incarnate for You: Conceptualizing God's Desire for the Flesh* (Lanham, Md.: Lexington, 2016). Some of what appears here can also be found formulated with different emphases in those works.

2. Lawrence Hass persuasively makes the case that there are three interrelated senses in which Merleau-Ponty uses the term "flesh" that need to be kept in mind and distinct when reading his work. See Lawrence Hass, *Merleau-Ponty's Philosophy* (Bloomington: Indiana University Press, 2008); and Adam Pryor, *Body of Christ Incarnate for You*, chap. 5. Here I am eliding features of the three senses of flesh to focus on the constructive potential of what Hass identifies as the least developed facet: flesh as an elemental ontology.

3. See Maurice Merleau-Ponty, *The Visible and the Invisible*, trans. Alphonso Lingis (Chicago: Northwestern University Press, 1968), 13.

4. Hass, *Merleau-Ponty's Philosophy*, 138.

5. For Merleau-Ponty what is common is a certain "style" of being, a form of expressive movement (most evident in perception) that characterizes the way in which different body-schemas make space for one another in the midst of the flesh (i.e., shared corporeality). On style, see Maurice Merleau-Ponty, *Phenomenology of Perception*, trans. Colin Smith (New York: Routledge & Kegan Paul, 1962), 378ff.; *Signs*, trans. Richard C. McCleary (Evanston: Northwestern University Press, 1964), 112ff.; and Scott L. Marratto, *The Intercorporeal Self: Merleau-Ponty on Subjectivity* (Albany: SUNY Press, 2012), 101–4 and 157–59.

6. See also Richard Kearney, *Anatheism (Returning to God after God)* (New York: Columbia University Press, 2010), 89–90.

7. Merleau-Ponty, *The Visible and the Invisible*, 147.

8. David Macauley, *Elemental Philosophy: Earth, Air, Fire, and Water as Environmental Ideas* (Albany: SUNY Press, 2010), 308. Citing Merleau-Ponty, *The Visible and the Invisible*, 147.

9. Glen A. Mazis, *Merleau-Ponty and the Face of the World: Silence, Ethics, Imagination, and Poetic Ontology* (Albany: SUNY Press, 2016),

265–56. This is not simply testifying to the long-held supposition that Being is only actualized through beings that are never in turn precisely equivalent to Being. Flesh is manifest in the contact *between* beings and the world. Flesh does not belong to *a* particular being but to an encounter.

10. Adam Pryor, "The Liminal as Kairos: A Cyborg Example for Theology and Science" in *The Many Voices of Liminality*, ed. Timothy Carson (Cambridge: Lutterworth Press: 2019).

11. See the examples in Merleau-Ponty, *Phenomenology of Perception*, 140–55; and *The Visible and the Invisible*, 147–48. I am taking these as the impetus for my own example of riding a bike.

12. Ted Toadvine, *Merleau-Ponty's Philosophy of Nature* (Evanston: Northwestern University Press, 2009), 110ff.

13. Merleau-Ponty, *The Visible and the Invisible*, 133. On this separation, see also Mazis, *Merleau-Ponty and the Face of the World*, 60–62; and Jean-Paul Sartre, *Being and Nothingness*, trans. Hazel E. Barnes, Reprint edition (New York: Washington Square Press, 1993), pt. 3, chap. 2.

14. The language here may seem strange. Our quotidian experience indicates that only something sentient does these things. I want to bracket that assumption and look holistically at the lived-experience of riding the bicycle. Even if we do not want to suggest that the bicycle senses me, in the perceptual encounter I can garner an awareness of how my lived-body can be sensed—indicating that this experience inheres in all flesh. On how one might thoroughly remove consciousness from consideration of a more primordial sense of lived-experience in the guise of Merlau-Ponty's ontology, see Renaud Barbaras in *The Being of the Phenomenon: Merleau-Ponty's Ontology*, trans. Ted Toadvine and Leonard Lawlor (Bloomington: Indiana University Press, 2004).

15. Merleau-Ponty, *The Visible and the Invisible*, 250.

16. Merleau-Ponty, *The Visible and the Invisible*, 135–56.

17. Hass, *Merleau-Ponty's Philosophy*, 128.

18. Mazis, *Merleau-Ponty and the Face of the World*, 266–67.

19. On the "crossed phenomenon," see Jean-Luc Marion, *The Erotic Phenomenon*, trans. Stephen E. Lewis (Chicago: University of Chicago Press, 2007), 105 and 189.

In turning to the crossed phenomenon, I am reading against Marion's own text in two critical senses. First, in his treatment of the flesh, Marion follows the work of Jacques Derrida and Jean-Luc Nancy. Each critiques Merleau-Ponty's flesh as if it re-inscribes a sense of the transcendental ego—as a self-justifying first metaphysical principle. Mayra Rivera and Scott Marratto, with whom my reading agrees, have each made rich cases against this line of critique directed at the flesh. Though an exhaustive treatment of this issue goes beyond our focus here, see *The Erotic Phenomenon*, 38–39 and 113–15; Jean-Luc Nancy, *Corpus*, trans. Richard Rand (New York: Fordham University Press, 2008), 47–88 and 122–35; Jacques

Derrida, *On Touching-Jean-Luc Nancy*, trans. Christine Irizarry (Stanford: Stanford University Press, 2005), 159–215; Marratto, *The Intercorporeal Self*, 128–41; Mayra Rivera, *Poetics of the Flesh* (Durham, N.C.: Duke University Press, 2015), 96–105; and Pryor, *Body of Christ Incarnate for You*, 79–87.

Second, any crossed phenomenon—like love—on Marion's account can only occur between two conscious entities. Technically put, I feel the feeling of the beloved's love for me as the beloved feels the feeling of my love for her. Only a beloved with the interiority of a mind could engage in this sort of encounter. I am not meaning to suggest that every instance of the flesh possesses some interiority or sense of mentality (one way by which we might widen the sphere of encounters to which the crossed phenomenon applies). Instead, I want to suggest that non-conscious entities cannot help but blindly be offered up in presence: They cannot help but risk the vulnerability of persistent exposure. This stands contra the "third passivity" in *The Erotic Phenomenon*, 110–12.

20. On the "erotic obstacle," see Marion, *The Erotic Phenomenon*, 44–47.

21. Marion, *The Erotic Phenomenon*, 106.

22. "Loving provisionally—this is nonsense, a contradiction in terms." Marion, *The Erotic Phenomenon*, 185.

23. Marion, *The Erotic Phenomenon*, 70–76.

24. Critical to Marion's argument is that this assurance of the meaning of being is more important than a mere assurance of bare existence. See *The Erotic Phenomenon*, 21–23.

25. Marion, *The Erotic Phenomenon*, 110.

26. On presence as openness, see Hans-Georg Gadamer, *Truth And Method*, trans. Joel Weinsheimer and Donald G. Marshall (New York: Crossroad Publishing, 1982), 110–11.

27. Sara Ahmed, *Queer Phenomenology: Orientations, Objects, Others* (Durham, N.C.: Duke University Press, 2006), 68. Ahmed departs from approaches that primarily use phenomenology to make a theoretical claim on the "everyday" lives of queer persons. For examples of queer phenomenology done in this vein or what it would mean to do queer phenomenology in this vein, see Lizabeth During and Terri Fealy, "Philosophy," in *Lesbian and Gay Studies: A Critical Introduction*, ed. Andy Medhurst and Sally Munt (London: Cassell, 1997); Henry S. Rubin, "Phenomenology as a Method in Trans Studies," *GLQ* 4, no. 2 (1998): 145–58; and Anabelle Willox, "Phenomenology, Embodiment and the Political Efficacy of Contingent Identity Claims," in *The Ashgate Research Companion to Queer Theory*, ed. Noreen Giffney and Michael O'Rourke, 1st edition (Burlington, Vt.: Routledge, 2009), 95–110.

28. Ahmed, *Queer Phenomenology*, 88. See also Don Ihde, *Technology and the Lifeworld: From Garden to Earth* (Bloomington: Indiana University Press, 1990).

29. Ahmed, *Queer Phenomenology*, 116.

30. Ahmed, *Queer Phenomenology*, 129. Phenomenologically, Ahmed's "orientation" is akin to "intentionality" as it relates to habit. Ahmed indicates this when connecting her idea of tendencies to *habitus*.

31. Within queer studies, there are many ways of accounting for the performativity of a "tending toward" producing naturalized "tendencies." Ahmed employs Adrienne Rich's concept of "compulsory heterosexuality," illustrating how wider discourse regarding orientation intersects with performativity that troubles the sex/gender divide. Ahmed, *Queer Phenomenology*, 84ff; Adrienne Rich, "Compulsory Heterosexuality and Lesbian Existence," in *The Lesbian and Gay Studies Reader*, ed. Henry Abelove, Michele Aina Barale, and David M. Halperin (New York: Routledge, 1993), 227–54.

32. See also Anthony J. Steinbock, *Home and Beyond: Generative Phenomenology after Husserl* (Evanston: Northwestern University Press, 1995), chaps. 1–2.

33. Ahmed, *Queer Phenomenology*, 54.

34. See Ahmed, *Queer Phenomenology*, 51–63.

35. To analytically divide this indirect experience of orientation in the flesh, we would have to step outside of the chiasmic experience itself and make this event the direct consideration of a different enfolding of the flesh, which would be characterized by its own separate, distinct, and indirect sense of orientation.

36. Ahmed, *Queer Phenomenology*, 52.

37. Receiving certain orientations and proximities does not necessarily imply we automatically reenact these orientations-around or orient-toward only those desires that are most proximate. There is performative disruption in our orientations: "*the gap between reception and possession.*" This gap opens a space for political resistance through re-orientation that shifts the proximities of future generations. Ahmed, *Queer Phenomenology*, 154.

38. Ahmed, *Queer Phenomenology*, 86; Judith Butler, "Performative Acts and Gender Constitutions: An Essay in Phenomenology and Feminist Theory," in *Writing on the Body: Female Embodiment and Feminist Theory*, ed. Katie Conboy, Nadia Medina, and Sarah Stanbury (New York: Columbia University Press, 1997).

39. Ahmed, *Queer Phenomenology*, 139.

40. Ahmed, *Queer Phenomenology*, 170–71.

41. Martin Heidegger, *Being and Time*, trans. John Macquarie and Edward Robinson, Reprint (New York: Harper Perennial Modern Classics, 2008), 232–34.

42. Ahmed, *Queer Phenomenology* 68–79. "Queer" clearly refers both to a general sense of being askew and non-normative sexualities, which "involves a personal and social commitment to living in an oblique world"

(161). The meanings are not synonymous but certainly related. I am aware that this use of "queer" and "queering" threatens to instrumentalize the experience of those living obliquely. This seems to be an inherent risk to any broadened use of the word "queer" or "queering" with regard to orientations besides sexuality that would move queer phenomenology out of its descriptive mode.

43. Ahmed, *Queer Phenomenology*, 171–72.

44. We are now getting a sense of why this approach is *phenomenologically* interesting. The bracketing of the natural attitude is a critical feature of phenomenologies. Queer phenomenology pays attention to the role of disorientation and queer phenomena in this *process of phenomenological bracketing.* A queer phenomenology is not one that simply starts from queer identity or queer experiences; it queers a critical feature of phenomenological thinking. While Ahmed turns more directly to imagining a queer politics, her account hints at the development of a *wondrous attitude* as an alternative to the phenomenological attitude and the phenomenological reduction's assumed bracketing. See Ahmed, *Queer Phenomenology*, 160 and 164.

45. Mary-Jane Rubenstein, *Strange Wonder: The Closure of Metaphysics and the Opening of Awe* (New York: Columbia University Press, 2008), 4.

46. Rubenstein, *Strange Wonder*, 10–11.

47. I am eliding some technical terms (particularly around anxiety, care, and *Veraltenheit*) that Rubenstein makes distinct. See *Strange Wonder*, 29, 36 and 59–60; and Martin Heidegger, *Basic Questions of Philosophy: Selected "Problems" of "Logic,"* trans. Andre Schuwer and Richard Rojcewicz (Bloomington: Indiana University Press, 1994), 142–44.

48. Heidegger, *Basic Questions of Philosophy*, 146–47.

49. Ahmed, *Queer Phenomenology*, 171.

50. Johan Huizinga, *Homo Ludens: A Study of the Play-Element in Culture* (Boston: Beacon Press, 1970), 45.

51. See Gadamer, *Truth And Method*, 92–93.

52. Courtney T. Goto, *The Grace of Playing: Pedagogies for Learning into God's New Creation* (Eugene, Ore.: Pickwick Publications, 2016), 15.

53. Robert N. Bellah, *Religion in Human Evolution: From the Paleolithic to the Axial Age* (Cambridge: Belknap Press, 2011), 2–4; Alfred Schutz, "Multiple Realities" (1945) in *Collected Papers*, vol. 1, *The Problem of Social Reality* (The Hague: Martinus Hijhoff, 1967), 207–59.

54. Gordon M. Burghardt, *The Genesis of Animal Play: Testing the Limits* (Cambridge: MIT Press, 2005), 71 and 77–78.

55. Huizinga, *Homo Ludens*, 8.

56. Goto, *The Grace of Playing*, 21–23.

57. Bellah, *Religion in Human Evolution*, xv.

58. Bellah, *Religion in Human Evolution*, xxi.

59. Gordon Burghardt's distinction between primary, secondary, and tertiary process play is helpful regarding this distinction. See Bellah, *Religion in Human Evolution*, xxii and 78–79; and Burghardt, *The Genesis of Animal Play*, 118–21.

60. Goto, *The Grace of Playing*, 16; Anna Maria Nicolò, "Playing," in *Playing and Reality Revisited: A New Look at Winnicott's Classic Work*, ed. Christian Seulin and Gennaro Saragnano (London: Karmac Books, 2015), 21–44.

61. Alison Gopnik, T*he Philosophical Baby: What Children's Minds Tell Us About Truth, Love, and the Meaning of Life* (New York: Farrar, Straus and Giroux, 2009), 19.

62. Burghardt, *The Genesis of Animal Play*, 74.

63. Huizinga, *Homo Ludens*, 13–15.

64. See also Goto, *The Grace of Playing*, 18–19.

65. Gopnik, *The Philosophical Baby*, 71–72.

66. Burghardt, *The Genesis of Animal Play*, 75.

67. Huizinga, *Homo Ludens*, 11. Compare to Goto, *The Grace of Playing*, 21–23.

68. I am implicitly relying on the analysis of Monica Vilhauer in her work *Gadamer's Ethics of Play: Hermeneutics and the Other* (Lanham, Md.: Lexington Books, 2010). She suggests that the play-process is a hermeneutical key for interpreting Gadamer's account of understanding, contending that a "fusion" of horizons often fails to sufficiently account for his dialogical interplay. Gadamer does not reference the flesh, but the ontological significance of play he envisions can create a link. See Gadamer, *Truth and Method*, 91–119.

69. Gadamer, *Truth and Method*, 93.

70. Gadamer, *Truth and Method*, 95–97.

71. At its best, our everyday experiences give us the deeply personal *Erlebnis* that might in their specific appeal to subjectivity be taken to stand as a fragment in which a greater sense of ontological wholeness is present, but never can everydayness give the transformative *Erfahrung*. See Gadamer, *Truth and Method*, 55–63 and 310–325; and Vilhauer, *Gadamer's Ethics of Play*, 47n15.

72. Gadamer, *Truth and Method*, 112–13.

73. This facilitates *Bildung* if we take the German philosophical tradition at face value. Gadamer, *Truth and Method*, 12–18.

74. Merleau-Ponty, *The Visible and the Invisible*, 93–95. The flesh understood this way is a product of Merlau-Ponty's hyperdialectic. See Pryor, *Body of Christ*, chaps. 5 and 9. Such an account fits neatly with Barad's understanding of intra-action found in *Meeting the Universe Halfway*.

75. I have in mind the sorts of indirect communication described by Søren Kierkegaard as a subjectivity of faith in *Practice in Christianity*,

trans. Howard Hong and Edna Hong (Princeton: Princeton University Press, 1991).

## Epilogue: *Ad Astra Per Aspera*

1. Natalie Thomas and Sumant Nigam, "Twentieth-Century Climate Change over Africa: Seasonal Hydroclimate Trends and Sahara Desert Expansion," *Journal of Climate* 31, no. 9 (March 28, 2018): 3349–70.

2. Vince Beiser, "China's Crazy Plan to Keep Sand From Swallowing the World," *Mother Jones*, October 2017, https://www.motherjones.com/environment/2017/08/china-plants-billions-of-trees-in-the-desert/.

3. Yan Li et al., "Climate Model Shows Large-Scale Wind and Solar Farms in the Sahara Increase Rain and Vegetation," *Science* 361, no. 6406 (September 7, 2018): 1019–22.

4. Ning Zeng and Jinho Yoon, "Expansion of the World's Deserts Due to Vegetation-Albedo Feedback under Global Warming," *Geophysical Research Letters* 36, no. 17 (September 1, 2009).

5. Dan Charles, "A Scientist Dreams Up a Plan to Stop the Sahara from Expanding," NPR, *Goats and Soda: Stories of Life in a Changing World* (blog), September 9, 2018, https://www.npr.org/sections/goatsandsoda/2018/09/09/645539064/so-maybe-stopping-the-sahara-from-expanding-isn-t-an-impossible-dream.

6. Xunming Wang et al., "Climate, Desertification, and the Rise and Collapse of China's Historical Dynasties," *Human Ecology* 38, no. 1 (2010): 157–72; Greger Larson et al., "Current Perspectives and the Future of Domestication Studies," *Proceedings of the National Academy of Sciences of the United States of America* 111, no. 17 (April 29, 2014): 6139–46; David K. Wright, "Humans as Agents in the Termination of the African Humid Period," *Frontiers in Earth Science* 5 (2017).

7. "Kansas Agriculture," Kansas Department of Agriculture, accessed September 19, 2018, http://agriculture.ks.gov/about-kda/kansas-agriculture.

8. "The Purpose of No-Till on the Plains," No-Till on the Plains: Agriculture Production Systems Modeling Nature, 2018, http://www.notill.org/about/purpose.

9. As the organization puts it in identifying their purpose, "Why would anyone want to accept a premise of sustaining degradation, which ultimately results in ruination of our resources and life?"

10. "Vision and Mission," The Land Institute, 2018, https://landinstitute.org/about-us/vision-mission/.

11. Timothy Crews et al., "New Roots for Ecological Intensification," *Crops, Soils, Agronomy News* 59, no. 11 (November 1, 2014): 16–17; Jerry D. Glover et al., "Harvested Perennial Grasslands Provide Ecological

Benchmarks for Agricultural Sustainability," *Agriculture, Ecosystems & Environment* 137, no. 1 (April 15, 2010): 3–12.

12. S. K. Kakraliya et al., "Performance of Portfolios of Climate Smart Agriculture Practices in a Rice-Wheat System of Western Indo-Gangetic Plains," *Agricultural Water Management* 202 (April 1, 2018): 122–33.

13. "Ecosphere Studies," The Land Institute, 2018, https://landinstitute.org/our-work/ecosphere-studies/.

14. "Mission and Vision," Lutherans Restoring Creation, February 28, 2018, https://lutheransrestoringcreation.org/about-us/vision/.

15. "Becoming a Creation-Care Congregation: A Self-Organizing Kit with Guidelines and Resources," June 2017, https://css-elca.nm-secure.com/files/congregational_self_organizing_kit_2017.pdf.

16. Shelley Wickstrom, "Introducing Shishmaref" (lecture, Convocation of the Association of Teaching Theologians, Minneapolis, July 31, 2018). I am deeply grateful to Shelley Wickstrom for providing me the text of her talk. The rich descriptions and many of the facts and figures come directly from her presentation.

17. Wickstrom, "Introducing Shishmaref."

# Bibliography

Ackerman, Frank, and Lisa Heinzerling. *Priceless: On Knowing the Price of Everything and the Value of Nothing.* New York: The New Press, 2005.

——. "Pricing the Priceless: Cost-Benefit Analysis of Environmental Protection." *University of Pennsylvania Law Review* 150, no. 5 (2002): 1553–84.

Adcock, C. T., E. M. Hausrath, and P. M. Forster. "Readily Available Phosphate from Minerals in Early Aqueous Environments on Mars." *Nature Geoscience* 6, no. 10 (October 2013): 824–27.

Ahmed, Sara. *Queer Phenomenology: Orientations, Objects, Others.* Durham, N.C.: Duke University Press, 2006.

Airapetian, Vladimir S., Alex Glocer, George V. Khazanov, R. O. P. Loyd, Kevin France, Jan Sojka, William C. Danchi, and Michael W. Liemohn. "How Hospitable Are Space Weather Affected Habitable Zones? The Role of Ion Escape." *The Astrophysical Journal Letters* 836, no. 1 (2017): L3.

Anglada-Escudé, Guillem, Pedro J. Amado, John Barnes, Zaira M. Berdiñas, R. Paul Butler, Gavin A. L. Coleman, Ignacio de la Cueva, et al. "A Terrestrial Planet Candidate in a Temperate Orbit around Proxima Centauri." *Nature* 536, no. 7617 (August 25, 2016): 437–40.

Atri, Dimitra. "Modelling Stellar Proton Event-Induced Particle Radiation Dose on Close-in Exoplanets." *Monthly Notices of the Royal Astronomical Society: Letters* 465, no. 1 (February 11, 2017): L34–38.

Augustine. *Confessions.* Translated by Henry Chadwick. Oxford: Oxford University Press, 1998.

——. *Eighty-Three Different Questions.* Translated by David Mosher. Washington, D.C: Catholic University of America Press, 1982.

——. *On Genesis.* Translated by Edmund Hill. Vol. 1:13. The Works of Saint Augustine. Hyde Park: New City Press, 2002.

———. *The Trinity*. Translated by Edmund Hill. Vol. 1:5. The Works of Saint Augustine. Brooklyn: New City Press, 1990.

Auld, Graeme. "Imago Dei in Genesis: Speaking in the Image of God." *The Expository Times* 116, no. 8 (May 2005): 259–62.

Austin, Jon. "Kepler-452b: How Long Would It Take Humans to Reach 'Earth 2' and Could We Live There?" Express.co.uk, July 28, 2015. http://www.express.co.uk/news/science/594133/Kepler-452b-How-long-take-humans-reach-Earth-2-could-we-live-there.

Bains, William. "Many Chemistries Could Be Used to Build Living Systems." *Astrobiology* 4, no. 2 (June 2004): 137–67.

Ballard, Sarah. "Predicted Number, Multiplicity, and Orbital Dynamics of TESS M Dwarf Exoplanets." *ArXiv:1801.04949 [Astro-Ph]*, January 15, 2018.

Barad, Karen. *Meeting the Universe Halfway: Quantum Physics and the Entanglement of Matter and Meaning*. Durham, N.C.: Duke University Press Books, 2007.

Barbaras, Renaud. *The Being of the Phenomenon: Merleau-Ponty's Ontology*. Translated by Ted Toadvine and Leonard Lawlor. Bloomington: Indiana University Press, 2004.

Barnes, Rory. "Opportunities and Obstacles for Life on Proxima B." *Pale Red Dot* (blog), August 28, 2016. https://palereddot.org/opportunities-and-obstacles-for-life-on-proxima-b/.

Barnes, Rory, Russell Deitrick, Rodrigo Luger, Peter E. Driscoll, Thomas R. Quinn, David P. Fleming, Benjamin Guyer, et al. "The Habitability of Proxima Centauri b I: Evolutionary Scenarios." *ArXiv:1608.06919 [Astro-Ph]*, August 24, 2016.

Barnosky, Anthony D., Nicholas Matzke, Susumu Tomiya, Guinevere O. U. Wogan, Brian Swartz, Tiago B. Quental, Charles Marshall, et al. "Has the Earth's Sixth Mass Extinction Already Arrived?" *Nature* 471, no. 7336 (March 3, 2011): 51–57.

Barth, Karl. *Church Dogmatics*. Edinburgh: T & T Clark, 1932–1967.

Baskin, Jeremy. "Paradigm Dressed as Epoch: The Ideology of the Anthropocene." *Environmental Values* 24, no. 1 (2015): 9–29.

Bauman, Whitney A. *Religion and Ecology: Developing a Planetary Ethic*. New York: Columbia University Press, 2014.

Beasley, Chris, and Carol Bacchi. "Envisaging a New Politics for an Ethical Future: Beyond Trust, Care and Generosity—Towards an Ethic of 'Social Flesh.'" *Feminist Theory* 8, no. 3 (December 1, 2007): 279–98.

"Becoming a Creation-Care Congregation: A Self-Organizing Kit with Guidelines and Resources," June 2017. https://css-elca.nm-secure.com/files/congregational_self_organizing_kit_2017.pdf.

Beiser, Vince. "China's Crazy Plan to Keep Sand From Swallowing the World." *Mother Jones*, October 2017. https://www.motherjones.com/environment/2017/08/china-plants-billions-of-trees-in-the-desert/.

Bellah, Robert N. *Religion in Human Evolution: From the Paleolithic to the Axial Age.* Cambridge: Belknap Press, 2011.

Benner, Steven A. "Defining Life." *Astrobiology* 10, no. 10 (December 2010): 1021–30.

Benner, Steven A., Alonso Ricardo, and Matthew A. Carrigan. "Is There a Common Chemical Model for Life in the Universe?" *Current Opinion in Chemical Biology* 8, no. 6 (December 2004): 672–89.

Bennett, Jane. *Vibrant Matter: A Political Ecology of Things.* Durham, N.C.: Duke University Press, 2010.

Berlejung, Angelika. *Die Theologie Der Bilder: Herstellung Und Einweihung von Kultbildern in Mesopotamien Und Die Alttestamentliche Bilderpolemik.* Fribourg: Vandenhoeck & Ruprecht, 1998.

Betcher, Sharon V. *Spirit and the Obligation of Social Flesh: A Secular Theology for the Global City.* New York: Fordham University Press, 2013.

Bibring, Jean-Pierre, Yves Langevin, John F. Mustard, François Poulet, Raymond Arvidson, Aline Gendrin, Brigitte Gondet, et al. "Global Mineralogical and Aqueous Mars History Derived from OMEGA/Mars Express Data." *Science* 312, no. 5772 (April 21, 2006): 400–4.

Boeke, Kees. *Cosmic View: The Universe in 40 Jumps.* New York: John Day Company, 1957.

Bonfils, X., N. Astudillo-Defru, R. Diaz, J.-M. Almenara, T. Forveille, F. Bouchy, X. Delfosse, et al. "A Temperate Exo-Earth Around a Quiet M Dwarf at 3.4 Parsecs." *Astronomy & Astrophysics*, November 15, 2017.

Bonneuil, Christophe. "The Geological Turn: Narratives of the Anthropocene." In *The Anthropocene and the Global Environmental Crisis: Rethinking Modernity in a New Epoch*, edited by Clive Hamilton, Christophe Bonneuil, and Francois Gemenne, 17–31. Abingdon: Routledge, 2015.

Bonneuil, Christophe, and Jean-Baptiste Fressoz. *The Shock of the Anthropocene: The Earth, History and Us.* London: Verso, 2016.

Braudel, Fernand. *The Mediterranean and the Mediterranean World in the Age of Philip II.* 2 vols. London: Fontana, 1966.

Brown, Delwin. "Public Theology, Academic Theology: Wentzel Van Huyssteen and the Nature of Theological Rationality." *American Journal of Theology & Philosophy* 22, no. 1 (2001): 88–101.

Brown, Robert H., Roger N. Clark, Bonnie J. Buratti, Dale P. Cruikshank, Jason W. Barnes, Rachel M. E. Mastrapa, J. Bauer, et al. "Composition and Physical Properties of Enceladus' Surface." *Science* 311, no. 5766 (March 10, 2006): 1425–28.

Brueggemann, Walter. *Genesis.* Interpretation. Atlanta: John Knox Press, 1982.

Brunner, Emil. *Man in Revolt: A Christian Anthropology.* Translated by Olive Wyon. Philadelphia: Westminster Press, 1947.

Burghardt, Gordon M. *The Genesis of Animal Play: Testing the Limits*. Cambridge: MIT Press, 2005.

Butler, Judith. "Performative Acts and Gender Constitutions: An Essay in Phenomenology and Feminist Theory." In *Writing on the Body: Female Embodiment and Feminist Theory*, edited by Katie Conboy, Nadia Medina, and Sarah Stanbury. New York: Columbia University Press, 1997.

Cairns, David. *The Image of God in Man*. London: Collins, 1973.

Carrington, Damian. "The Anthropocene Epoch: Scientists Declare Dawn of Human-Influenced Age." *The Guardian*, August 29, 2016, sec. Science. https://www.theguardian.com/environment/2016/aug/29/declare -anthropocene-epoch-experts-urge-geological-congress-human-impact -earth.

Carter, J., F. Poulet, J.-P. Bibring, N. Mangold, and S. Murchie. "Hydrous Minerals on Mars as Seen by the CRISM and OMEGA Imaging Spectrometers: Updated Global View." *Journal of Geophysical Research: Planets* 118, no. 4 (April 1, 2013): 831–58.

Caseldine, Chris. "Conceptions of Time in (Paleo) Climate Science and Some Implications." *Wiley Interdisciplinary Reviews: Climate Change* 3, no. 4 (2012): 329–38.

Catling, David C. *Astrobiology: A Very Short Introduction*. Oxford: Oxford University Press, 2013.

Chakrabarty, Dipesh. "The Climate of History: Four Theses." *Critical Inquiry* 35, no. 2 (2009): 197–222.

Chang, Kenneth. "One Star Over, a Planet That Might Be Another Earth." *The New York Times*, August 24, 2016. https://www.nytimes.com/ 2016/08/25/science/earth-planet-proxima-centauri.html.

Chardin, Pierre Teilhard de. *The Phenomenon of Man*. Translated by Bernard Wall. New York: Harper Perennial, 1976.

Charles, Dan. "A Scientist Dreams Up a Plan to Stop the Sahara from Expanding." NPR. *Goats and Soda: Stories of Life in a Changing World* (blog), September 9, 2018. https://www.npr.org/sections/goatsandsoda/ 2018/09/09/645539064/so-maybe-stopping-the-sahara-from-expanding -isn-t-an-impossible-dream.

Chen, Jingjing, and David Kipping. "Probabilistic Forecasting of the Masses and Radii of Other Worlds." *The Astrophysical Journal* 834, no. 1 (2017): 17.

Choi, Charles Q. "Mars Life? 20 Years Later, Debate Over Meteorite Continues." Space.com. Accessed July 12, 2018. https://www.space .com/33690-allen-hills-mars-meteorite-alien-life-20-years.html.

Clingerman, Forrest. "Place and the Hermeneutics of the Anthropocene." *Worldviews: Global Religions, Culture, and Ecology* 20, no. 3 (January 1, 2016): 225–37.

Clingerman, Forrest, Brian Treanor, Martin Drenthen, and David Utsler, eds. *Interpreting Nature: The Emerging Field of Environmental Hermeneutics*. New York: Fordham University Press, 2013.

Clough, David L. *On Animals: Volume I: Systematic Theology*. Edinburgh: T & T Clark, 2014.

Cockell, C. S., T. Bush, C. Bryce, S. Direito, M. Fox-Powell, J. P. Harrison, H. Lammer, et al. "Habitability: A Review." *Astrobiology* 16, no. 1 (January 2016): 89–117.

Cousins, Ewert. *Bonaventure and the Coincidence of Opposites*. Chicago: Franciscan Herald Press, 1978.

Crews, Timothy, Thomas Cox, Lee De Haan, Sivaramakrishna Damaraju, Wes Jackson, Pheonah Nabukalu, David Van Tassel, and Shuwen Wang. "New Roots for Ecological Intensification." *Crops, Soils, Agronomy News* 59, no. 11 (November 1, 2014): 16–17.

Crist, Eileen. "On the Poverty of Our Nomenclature." *Environmental Humanities* 3 (2013): 129–47.

Crockett, Clayton. *Deleuze Beyond Badiou: Ontology, Multiplicity, and Event*. New York: Columbia University Press, 2013.

Crutzen, Paul J. "Geology of Mankind." *Nature* 415 (January 3, 2002): 23.

Crutzen, Paul J., and Christian Schwägerl. "Living in the Anthropocene: Toward a New Global Ethos." *Yale Environment 360*, January 24, 2011.

Damon, Maria, Kristina Mohlin, and Thomas Sterner. "Putting a Price on the Future of Our Children and Grandchildren." In *The Globalization of Cost-Benefit Analysis in Environmental Policy*, edited by Michael A. Livermore and Richard L. Revesz, chapter 4. Oxford: Oxford University Press, 2013.

Davies, Jeremy. *The Birth of the Anthropocene*. Oakland: University of California Press, 2016.

Deamer, David W., and Gail R. Fleischaker. *Origins of Life: The Central Concepts*. Boston: Jones & Bartlett Learning, 1994.

Deane-Drummond, Celia E., and David Clough. *Creaturely Theology: God, Humans and Other Animals*. London: Hymns Ancient & Modern Ltd, 2009.

Deane-Drummond, Celia. "God's Image and Likeness in Humans and Other Animals: Performative Soul-Making and Graced Nature." *Zygon: Journal of Religion and Science* 47, no. 4 (December 2012): 934–48.

Deleuze, Gilles. *Difference and Repetition*. Translated by Paul Patton. New York: Columbia University Press, 1995.

Derrida, Jacques. *On Touching-Jean-Luc Nancy*. Translated by Christine Irizarry. Stanford: Stanford University Press, 2005.

Di Achille, Gaetano, and Brian M. Hynek. "Ancient Ocean on Mars Supported by Global Distribution of Deltas and Valleys." *Nature Geoscience* 3, no. 7 (July 2010): 459–63.

Dibley, Ben. "'Nature Is Us': The Anthropocene and Species-Being." *Transformations* 21 (2012).

Domagal-Goldman, Shawn D., Katherine E. Wright, Katarzyna Adamala, Leigh Arina de la Rubia, Jade Bond, Lewis R. Dartnell, Aaron D. Goldman, et al. "The Astrobiology Primer v2.0." *Astrobiology* 16, no. 8 (August 1, 2016): 561–653.

Driscoll, Peter E., and Rory Barnes. "Tidal Heating of Earth-like Exoplanets around M Stars: Thermal, Magnetic, and Orbital Evolutions." *Astrobiology* 15, no. 9 (September 1, 2015): 739–60.

During, Lizabeth, and Terri Fealy. "Philosophy." In *Lesbian and Gay Studies: A Critical Introduction*, edited by Andy Medhurst and Sally Munt. London: Cassell, 1997.

Eames, Charles, and Ray Eames. *The Powers of Ten.* Short Documentary. IBM, 1977.

Earle, Steven. *Physical Geology.* 1.5. BC Open Textbook Project, 2015. https://opentextbc.ca/geology/.

Echeverría Falla, Cecilia. "La Imagen de Dios En El Hombre: Consideraciones En Torno a La Cuestión En W. Pannenberg." *Scripta Theologica* 45, no. 3 (December 2013): 737–55.

Edwards, Lin. "Ikaros Unfurls First Ever Solar Sail in Space." PhysOrg.com, June 11, 2010. https://phys.org/news/2010-06-ikaros-unfurls-solar -space.html.

Eigenbrode, Jennifer L., Roger E. Summons, Andrew Steele, Caroline Freissinet, Maëva Millan, Rafael Navarro-González, Brad Sutter, et al. "Organic Matter Preserved in 3-Billion-Year-Old Mudstones at Gale Crater, Mars." *Science* 360, no. 6393 (June 8, 2018): 1096–1101.

Farley, Edward. *Deep Symbols: Their Postmodern Effacement and Reclamation.* Valley Forge, Pa.: Trinity Press International, 1996.

Fastook, James L., and James W. Head. "Glaciation in the Late Noachian Icy Highlands: Ice Accumulation, Distribution, Flow Rates, Basal Melting, and Top-down Melting Rates and Patterns." *Planetary and Space Science* 106 (2015): 82–98.

Fergusson, David. "Humans Created According to the Imago Dei: An Alternative Proposal." *Zygon: Journal of Religion and Science* 48, no. 2 (May 2013): 439–53.

Finley, Thomas. "Dimensions of the Hebrew Word for 'Create' (Bara')." *Bibliotheca Sacra* 148, no. 592 (October 1991): 409–23.

Flanagan, Thomas. "The Agricultural Argument and Original Appropriation: Indian Lands and Political Philosophy." *Canadian Journal of Political Science* 22, no. 3 (September 1989): 589–602.

Frank, Adam. "Climate Change and the Astrobiology of the Anthropocene." 13.7 Cosmos & Culture: Commentary on Science and Society, October 1, 2016. http://www.npr.org/sections/13.7/2016/10/01/ 495437158/climate-change-and-the-astrobiology-of-the-anthropocene.

———. *Light of the Stars: Alien Worlds and the Fate of the Earth.* New York: Norton, 2018.

Frank, Adam, and Woodruff Sullivan. "A New Empirical Constraint on the Prevalence of Technological Species in the Universe." *Astrobiology* 16, no. 5 (April 22, 2016): 359–62.

———. "Sustainability and the Astrobiological Perspective: Framing Human Futures in a Planetary Context." *Anthropocene* 5 (March 1, 2014): 32–41.

Franklin, S. "Science as Culture, Cultures as Science." *Annual Review of Anthropology* 24 (1995): 163–84.

Fuentes, Agustin. "Evolutionary Perspectives and Transdisciplinary Intersections: A Roadmap to Generative Areas of Overlap in Discussing Human Nature." *Theology and Science* 11, no. 2 (May 2013): 106–29.

Gadamer, Hans-Georg. *Truth And Method.* Translated by Joel Weinsheimer and Donald G. Marshall. New York: Crossroad Publishing, 1982.

Gannon, Megan. "Aliens Next Door: Does Proxima b Host Life?" Space. com, August 24, 2016. http://www.space.com/33846-proxima-b-alien -life-hunt.html.

Garcia-Ruiz, Juan-Manuel. "Morphological Behavior of Inorganic Precipitation Systems." In *SPIE Proceedings* 3755 (1999): 74–82.

Georges Louis Leclerc Buffon. *Les époques de la nature.* Paris: De l'imprimerie royale, 1780.

Ghosh, Amitav. *The Great Derangement.* Chicago: University of Chicago Press, 2016.

Gillon, Michaël, Amaury H. M. J. Triaud, Brice-Olivier Demory, Emmanuël Jehin, Eric Agol, Katherine M. Deck, Susan M. Lederer, et al. "Seven Temperate Terrestrial Planets Around the Nearby Ultracool Dwarf Star Trappist-1." *Nature* 542, no. 7642 (February 2017): 456–60.

Giri, A. K. "The Calling of a Creative Transdisciplinarity." *Futures* 34 (2002): 103–15.

Glein, Christopher R., John A. Baross, and J. Hunter Waite Jr. "The PH of Enceladus' Ocean." *Geochimica et Cosmochimica Acta* 162 (August 1, 2015): 202–19.

Glover, Jerry D., Steve W. Culman, S. Tianna DuPont, Whitney Broussard, Lauren Young, Margaret E. Mangan, John G. Mai, et al. "Harvested Perennial Grasslands Provide Ecological Benchmarks for Agricultural Sustainability." *Agriculture, Ecosystems & Environment* 137, no. 1 (April 15, 2010): 3–12.

Gopnik, Alison. *The Philosophical Baby: What Children's Minds Tell Us About Truth, Love, and the Meaning of Life.* New York: Farrar, Straus and Giroux, 2009.

Goto, Courtney T. *The Grace of Playing: Pedagogies for Learning into God's New Creation.* Eugene, Oregon: Pickwick Publications, 2016.

Grinspoon, David. *Earth in Human Hands: Shaping Our Planet's Future.* New York: Grand Central Publishing, 2016.

Guinan, E. F., and N. D. Morgan. "Proxima Centauri: Rotation, Chromospheric Activity, and Flares." *Bulletin of the American Astronomical Society* 28 (May 1996): 942.

Haff, P. K. "Technology as a Geological Phenomenon: Implications for Human Well-Being." *Geological Society of London Special Publications* 395 (May 1, 2014): 301–9.

Haff, Peter. "Humans and Technology in the Anthropocene: Six Rules." *The Anthropocene Review* 1, no. 2 (August 1, 2014): 126–36.

Haraway, Donna. *Simians, Cyborgs, and Women: The Reinvention of Nature.* New York: Routledge, 1991.

———. *Staying with the Trouble: Making Kin in the Chthulucene.* Durham, N.C.: Duke University Press, 2016.

Hartmann, William K., and Gerhard Neukum. "Cratering Chronology and the Evolution of Mars." *Space Science Reviews* 96 (2001): 165–94.

Hass, Lawrence. *Merleau-Ponty's Philosophy.* Bloomington: Indiana University Press, 2008.

Hays, Lindsay E., ed. "Astrobiology Strategy." NASA, October 2015. https://astrobiology.nasa.gov/uploads/filer_public/01/28/01283266-e401 -4dcb-8e05-3918b21edb79/nasa_astrobiology_strategy_2015_151008 .pdf.

Heath, M. J., L. R. Doyle, M. M. Joshi, and R. M. Haberle. "Habitability of Planets around Red Dwarf Stars." *Origins of Life and Evolution of the Biosphere: The Journal of the International Society for the Study of the Origin of Life* 29, no. 4 (August 1999): 405–24.

Hedman, M. M., C. M. Gosmeyer, P. D. Nicholson, C. Sotin, R. H. Brown, R. N. Clark, K. H. Baines, B. J. Buratti, and M. R. Showalter. "An Observed Correlation Between Plume Activity and Tidal Stresses on Enceladus." *Nature* 500, no. 7461 (August 8, 2013): 182–84.

Hefner, Philip. *The Human Factor.* Minneapolis: Augsburg Fortress Publishers, 2000.

Heidegger, Martin. *Basic Questions of Philosophy: Selected "Problems" of "Logic."* Translated by Andre Schuwer and Richard Rojcewicz. Bloomington: Indiana University Press, 1994.

———. *Being and Time.* Translated by John Macquarie and Edward Robinson. Reprint. New York: Harper Perennial Modern Classics, 2008.

Herzfeld, Noreen. *In Our Image: Artificial Intelligence and the Human Spirit.* Minneapolis: Fortress Press, 2002.

———. *Technology and Religion: Remaining Human in a Co-Created World.* Philadelphia: Templeton Press, 2009.

Hirsch Hadorn, Gertrude, David Bradley, Christian Pohl, Stephan Rist, and Urs Wiesmann. "Implications of Transdisciplinarity for Sustainability Research." *Ecological Economics* 60, no. 1 (November 1, 2006): 119–28.

Hoehler, Tori M. "An Energy Balance Concept for Habitability." *Astrobiology* 7, no. 6 (December 2007): 824–38.

Hoehler, Tori M., Jan P. Amend, and Everett L. Shock. "A 'Follow the Energy' Approach for Astrobiology." *Astrobiology* 7, no. 6 (December 1, 2007): 819–23.

Holm, N. G., C. Oze, O. Mousis, J. H. Waite, and A. Guilbert-Lepoutre. "Serpentinization and the Formation of $H_2$ and $CH_4$ on Celestial Bodies (Planets, Moons, Comets)." *Astrobiology* 15, no. 7 (July 1, 2015): 587–600.

Horlick-Jones, Tom, and Jonathan Sime. "Living on the Border: Knowledge, Risk and Transdisciplinarity." *Futures* 36, no. 4 (May 1, 2004): 441–56.

Hsu, Hsiang-Wen, Frank Postberg, Yasuhito Sekine, Takazo Shibuya, Sascha Kempf, Mihály Horányi, Antal Juhász, et al. "Ongoing Hydrothermal Activities Within Enceladus." *Nature* 519, no. 7542 (March 12, 2015): 207–10.

Huizinga, Johan. *Homo Ludens: A Study of the Play-Element in Culture.* Boston: Beacon Press, 1970.

Husserl, Edmund. *Formal Und Transzendentale Logik. Versuch Einer Kritik Der Logischen Vernunft. Mit Ergäzenden Texten.* Edited by Paul Janssen. Vol. 17. Husserliana. The Hague: Martinus Nijhoff, 1974.

Huyssteen, Wentzel van. *Alone in the World?: Human Uniqueness in Science and Theology.* Grand Rapids: Eerdmans, 2006.

———. *The Shaping of Rationality: Toward Interdisciplinarity in Theology and Science.* Grand Rapids: Eerdmans, 1999.

———. "What Makes Us Human? The Interdisciplinary Challenge to Theological Anthropology and Christology." *Toronto Journal of Theology* 26, no. 2 (September 1, 2010): 143–60.

Iess, L., D. J. Stevenson, M. Parisi, D. Hemingway, R. A. Jacobson, J. I. Lunine, F. Nimmo, et al. "The Gravity Field and Interior Structure of Enceladus." *Science* 344, no. 6179 (April 4, 2014): 78–80.

Ihde, Don. *Technology and the Lifeworld: From Garden to Earth.* Bloomington: Indiana University Press, 1990.

Irenaeus. *Saint Irenaeus of Lyons: Against Heresies.* Edited by Alexander Roberts, James Donaldson, and A. Cleveland Coxe. Indiana: Ex Fontibus, 2010.

Jakosky, Bruce. *Science, Society, and the Search for Life in the Universe.* Tucson: University of Arizona Press, 2006.

Jakosky, Bruce M., Joseph M. Grebowsky, Janet G. Luhmann, and David A. Brain. "Initial Results from the MAVEN Mission to Mars." *Geophysical Research Letters* 42, no. 21 (November 16, 2015): 2015GL065271.

Jenkins, Jon M., Joseph D. Twicken, Natalie M. Batalha, Douglas A. Caldwell, William D. Cochran, Michael Endl, David W. Latham, et al. "Discovery and Validation of Kepler-452b: A 1.6 $R_\oplus$ Super Earth Exoplanet in the Habitable Zone of a G2 Star." *The Astronomical Journal* 150, no. 2 (2015): 56.

Jia, Xianzhe, Margaret G. Kivelson, Krishan K. Khurana, and William S. Kurth. "Evidence of a Plume on Europa from Galileo Magnetic and Plasma Wave Signatures." *Nature Astronomy* (May 14, 2018): 1.

Jones, Gwyneth. *Proof of Concept.* Tor.com, 2017.

Jong, Jonathan, and Aku Visala. "Three Quests for Human Nature: Some Philosophical Reflections." *Philosophy, Theology, and the Sciences* 1, no. 2 (2014): 146–71.

"Kansas Agriculture," Kansas Department of Agriculture, http:// agriculture.ks.gov/about-kda/kansas-agriculture.

Kakraliya, S. K., H. S. Jat, Ishwar Singh, Tek B. Sapkota, Love K. Singh, Jhabar M. Sutaliya, Parbodh C. Sharma, et al. "Performance of Portfolios of Climate Smart Agriculture Practices in a Rice-Wheat System of Western Indo-Gangetic Plains." *Agricultural Water Management* 202 (April 1, 2018): 122–33.

Käsemann, Ernst. *Perspectives on Paul*. Translated by Margaret Kohl. Philadelphia: Fortress Press, 1971.

Kasting, J. F. "Runaway and Moist Greenhouse Atmospheres and the Evolution of Earth and Venus." *Icarus* 74 (June 1, 1988): 472–94.

Kasting, J. F., D. P. Whitmire, and R. T. Reynolds. "Habitable Zones around Main Sequence Stars." *Icarus* 101 (1993): 108–28.

Kaufman, Gordon D. *An Essay on Theological Method*. 3rd ed. Atlanta: Oxford University Press, 1995.

Kearney, Richard. *Anatheism (Returning to God after God)*. New York: Columbia University Press, 2010.

Keller, Catherine. *Cloud of the Impossible: Negative Theology and Planetary Entanglement*. New York: Columbia University Press, 2014.

———. *Face of the Deep: A Theology of Becoming*. New York: Routledge, 2002.

Keller, Catherine, and Mary-Jane Rubenstein, eds. *Entangled Worlds: Religion, Science, and New Materialisms*. New York: Fordham University Press, 2017.

Kelley, Deborah S., Jeffrey A. Karson, Gretchen L. Früh-Green, Dana R. Yoerger, Timothy M. Shank, David A. Butterfield, John M. Hayes, et al. "A Serpentinite-Hosted Ecosystem: The Lost City Hydrothermal Field." *Science* 307, no. 5714 (March 4, 2005): 1428–34.

Kierkegaard, Søren. *The Concept of Anxiety: A Simple Psychologically Orienting Deliberation on the Dogmatic Issue of Hereditary Sin*. Translated by Reidar Thomte and Albert B. Anderson. Princeton: Princeton University Press, 1981.

———. *Practice in Christianity*. Translated by Howard Hong and Edna Hong. Princeton: Princeton University Press, 1991.

Kipping, David M., Chris Cameron, Joel D. Hartman, James R. A. Davenport, Jaymie M. Matthews, Dimitar Sasselov, Jason Rowe, et al. "No Conclusive Evidence for Transits of Proxima b in MOST Photometry." *The Astronomical Journal* 153, no. 3 (February 2, 2017): 93.

Klein, J. T. *Crossing Boundaries: Knowledge, Disciplines, and Interdisciplinarities*. Charlottesville: University of Virginia Press, 1996.

Kohák, Erazim V. *The Embers and the Stars: A Philosophical Inquiry into the Moral Sense of Nature*. Chicago: University of Chicago Press, 1984.

Kolbert, Elizabeth. *The Sixth Extinction: An Unnatural History*. New York: Picador, 2015.

Kopparapu, Ravi Kumar, Ramses Ramirez, James F. Kasting, Vincent Eymet, Tyler D. Robinson, Suvrath Mahadevan, Ryan C. Terrien, Shawn Domagal-Goldman, Victoria Meadows, and Rohit Deshpande. "Habitable Zones Around Main-Sequence Stars: New Estimates." *The Astrophysical Journal* 765, no. 2 (March 10, 2013): 131.

Kretzmann, Norman, and Eleonore Stump, eds. *The Cambridge Companion to Augustine*. Cambridge: Cambridge University Press, 2001.

Krissansen-Totton, Joshua, and David C. Catling. "Constraining Climate Sensitivity and Continental Versus Seafloor Weathering Using an Inverse Geological Carbon Cycle Model." *Nature Communications* 8 (May 22, 2017): ncomms15423.

Kull, Anne. "Speaking Cyborg: Technoculture and Technonature." *Zygon* 37, no. 2 (June 2002): 279–87.

Lancaster, Sarah Heaner. "Divine Relations of the Trinity: Augustine's Answer to Arianism." *Calvin Theological Journal* 34, no. 2 (November 1, 1999): 327–46.

Larson, Greger, Dolores R. Piperno, Robin G. Allaby, Michael D. Purugganan, Leif Andersson, Manuel Arroyo-Kalin, Loukas Barton, et al. "Current Perspectives and the Future of Domestication Studies." *Proceedings of the National Academy of Sciences of the United States of America* 111, no. 17 (April 29, 2014): 6139–46.

Latour, Bruno. *Politics of Nature: How to Bring the Sciences into Democracy*. Translated by Catherine Porter. Cambridge: Harvard University Press, 2004.

Lawson, John. *The Biblical Theology of Saint Irenaeus*. London: Epworth Press, 1948.

Lengwiler, Martin, Michael Guggenheim, and Sabine Maasen. "Between Charisma and Heuristics: Interdisciplinary Practices in Applied Research Projects." *Science and Public Policy* 33, no. 6 (2006): 423–34.

Leopold, Aldo. *A Sand County Almanac, and Sketches Here and There*. Oxford: Oxford University Press, 1989.

Lewis, C. S. *Studies in Words*. Cambridge: Cambridge University Press, 1960.

Li, Yan, Eugenia Kalnay, Safa Motesharrei, Jorge Rivas, Fred Kucharski, Daniel Kirk-Davidoff, Eviatar Bach, and Ning Zeng. "Climate Model Shows Large-Scale Wind and Solar Farms in the Sahara Increase Rain and Vegetation." *Science* 361, no. 6406 (September 7, 2018): 1019–22.

Link, Lindsey S., Bruce M. Jakosky, and Geoffrey D. Thyne. "Biological Potential of Low-Temperature Aqueous Environments on Mars." *International Journal of Astrobiology* 4, no. 2 (April 2005): 155–64.

Linzey, Andrew. *Animal Theology*. Urbana: University of Illinois Press, 1995.

Lovelock, J. E. "Gaia as Seen Through the Atmosphere." *Atmospheric Environment (1967)* 6, no. 8 (August 1, 1972): 579–80.

Lovelock, James E., and Lynn Margulis. "Atmospheric Homeostasis by and for the Biosphere: The Gaia Hypothesis." *Tellus* 26, no. 1–2 (January 1, 1974): 2–10.

Lovin, Robin W., and Joshua Mauldin, eds. *Theology as Interdisciplinary Inquiry: Learning with and from the Natural and Human Sciences.* Grand Rapids: Eerdmans, 2017.

Lowenthal, David. *George Perkins Marsh: Prophet of Conservation.* Seattle: University of Washington Press, 2009.

———. "Nature and Morality from George Perkins Marsh to the Millennium." *Journal of Historical Geography* 26, no. 1 (January 1, 2000): 3–23.

Lubin, P. "A Roadmap to Interstellar Flight." *Journal of the British Interplanetary Society* 69 (2016): 40–72.

Luger, Rodrigo, and Rory Barnes. "Extreme Water Loss and Abiotic $O_2$ Buildup on Planets Throughout the Habitable Zones of M Dwarfs." *Astrobiology* 15, no. 2 (January 28, 2015): 119–43.

Macauley, David. *Elemental Philosophy: Earth, Air, Fire, and Water as Environmental Ideas.* Albany: SUNY Press, 2010.

MacKenzie, Shannon M., Tess E. Caswell, Charity M. Phillips-Lander, E. Natasha Stavros, Jason D. Hofgartner, Vivian Z. Sun, Kathryn E. Powell, et al. "THEO Concept Mission: Testing the Habitability of Enceladus's Ocean." *Advances in Space Research* 58, no. 6 (September 2016): 1117–37.

Malm, Andreas, and Alf Hornborg. "The Geology of Mankind? A Critique of the Anthropocene Narrative." *Anthropocene Review* 1, no. 1 (2014): 62–69.

Marion, Jean-Luc. *The Erotic Phenomenon.* Translated by Stephen E. Lewis. Chicago: University of Chicago Press, 2007.

Marratto, Scott L. *The Intercorporeal Self: Merleau-Ponty on Subjectivity.* Albany: SUNY Press, 2012.

"Mars Ice Deposit Holds as Much Water as Lake Superior." NASA/JPL, November 22, 2016.

Martin, Andrew, and Andrew McMinn. "Sea Ice, Extremophiles and Life on Extra-Terrestrial Ocean Worlds." *International Journal of Astrobiology* 17, no. 1 (January 2018): 1–16.

Martin, William, John Baross, Deborah Kelley, and Michael J. Russell. "Hydrothermal Vents and the Origin of Life." *Nature Reviews Microbiology* 6, no. 11 (November 2008): 805–14.

Martin, William, and Michael J. Russell. "On the Origins of Cells: A Hypothesis for the Evolutionary Transitions from Abiotic Geochemistry to Chemoautotrophic Prokaryotes, and from Prokaryotes to Nucleated Cells." *Philosophical Transactions of the Royal Society B: Biological Sciences* 358, no. 1429 (January 29, 2003): 59–85.

Martín-Torres, F. Javier, María-Paz Zorzano, Patricia Valentín-Serrano, Ari-Matti Harri, Maria Genzer, Osku Kemppinen, Edgard G. Rivera-Valentin, et al. "Transient Liquid Water and Water Activity at Gale Crater on Mars." *Nature Geoscience* 8, no. 5 (May 2015): 357–61.

Matsubara, Yo, Alan D. Howard, and J. Parker Gochenour. "Hydrology of Early Mars: Valley Network Incision." *Journal of Geophysical Research: Planets* 118, no. 6 (June 1, 2013): 1365–87.

Mays, James Luther. *Psalms.* Interpretation. Louisville: Westminster John Knox Press, 1994.

Mazis, Glen A. *Merleau-Ponty and the Face of the World: Silence, Ethics, Imagination, and Poetic Ontology.* Albany: SUNY Press, 2016.

McFadyen, Alistair. "Imaging God: A Theological Answer to the Anthropological Question?" *Zygon: Journal of Religion and Science* 47, no. 4 (December 2012): 918–33.

McFague, Sallie. *Metaphorical Theology: Models of God in Religious Language.* Philadelphia: Fortress Press, 1982.

McKay, Christopher P., Carolyn C. Porco, Travis Altheide, Wanda L. Davis, and Timothy A. Kral. "The Possible Origin and Persistence of Life on Enceladus and Detection of Biomarkers in the Plume." *Astrobiology* 8, no. 5 (October 2008): 909–19.

McKay, David S., Everett K. Gibson, Kathie L. Thomas-Keprta, Hojatollah Vali, Christopher S. Romanek, Simon J. Clemett, Xavier D. F. Chillier, Claude R. Maechling, and Richard N. Zare. "Search for Past Life on Mars: Possible Relic Biogenic Activity in Martian Meteorite ALH84001." *Science* 273, no. 5277 (August 16, 1996): 924–30.

McKay, Don. "Ediacaran and Anthropocene: Poetry as a Reader for Deep Time." In *Making the Geologic Now*, edited by Elizabeth Ellsworth and Jamie Kruse, 46–54. New York: Punctum, 2013.

McNulty, Michael T. "Augustine's Argument for the Existence of Other Souls." *Modern Schoolman: A Quarterly Journal of Philosophy* 48 (November 1, 1970): 19–24.

Melanchthon, Philipp. *A Melanchthon Reader.* Translated by Ralph Keen. New York: P. Lang, 1988.

Merleau-Ponty, Maurice. *Phenomenology of Perception.* Translated by Colin Smith. New York: Routledge & Kegan Paul, 1962.

———. *Signs.* Translated by Richard C. McCleary. Evanston: Northwestern University Press, 1964.

———. *The Visible and the Invisible.* Translated by Alphonso Lingis. Chicago: Northwestern University Press, 1968.

Mickol, R. L., and T. A. Kral. "Low Pressure Tolerance by Methanogens in an Aqueous Environment: Implications for Subsurface Life on Mars." *Origins of Life and Evolution of Biospheres*, September 23, 2016, 1–22.

Middleton, J. Richard. *The Liberating Image: The Imago Dei in Genesis 1.* Grand Rapids: Brazos Press, 2005.

Millard, A. R., and P. Bordreuil. "A Statue from Syria with Assyrian and Aramaic Inscriptions." *The Biblical Archeologist* 45 (1982): 135–41.

Milliken, R. E., G. A. Swayze, R. E. Arvidson, J. L. Bishop, R. N. Clark, B. L. Ehlmann, R. O. Green, et al. "Opaline Silica in Young Deposits on Mars." *Geology* 36, no. 11 (November 1, 2008): 847–50.

Minns, D. "Irenaeus." *Expository Times* 120, no. 4 (2009): 157–66.

"Mission and Vision." Lutherans Restoring Creation, February 28, 2018. https://lutheransrestoringcreation.org/about-us/vision/.

Mix, Lucas John. *Life in Space: Astrobiology for Everyone*. Cambridge: Harvard University Press, 2009.

Möhlmann, Diedrich T. F. "Are Nanometric Films of Liquid Undercooled Interfacial Water Bio-Relevant?" *Cryobiology* 58, no. 3 (June 2009): 256–61.

———. "Water in the Upper Martian Surface at Mid- and Low-Latitudes: Presence, State, and Consequences." *Icarus* 168, no. 2 (April 2004): 318–23.

Monastersky, Richard. "Anthropocene: The Human Age." *Nature* 519, no. 7542 (March 12, 2015): 144–47.

Moore, Stephen D., and Mayra Rivera, eds. *Planetary Loves: Spivak, Postcoloniality, and Theology*. New York: Fordham University Press, 2010.

Moritz, Joshua M. "Human Uniqueness, the Other Hominids, and 'Anthropocentrism of the Gaps' in the Religion and Science Dialogue." *Zygon: Journal of Religion and Science* 47, no. 1 (March 2012): 65–96.

Morton, Timothy. *Ecology without Nature: Rethinking Environmental Aesthetics*. Cambridge: Harvard University Press, 2009.

Mullally, Fergal, Susan E. Thompson, Jeffery L. Coughlin, Christopher J. Burke, and Jason F. Rowe. "Kepler's Earth-like Planets Should Not Be Confirmed Without Independent Detection: The Case of Kepler-452b." *The Astronomical Journal* 155, no. 5 (April 25, 2018): 210.

Nancy, Jean-Luc. *Corpus*. Translated by Richard Rand. New York: Fordham University Press, 2008.

"Nation Demands NASA Stop Holding Press Conferences Until They Discover Some Little Alien Guys." The Onion, September 28, 2015. https://www.theonion.com/nation-demands-nasa-stop-holding-press-conferences-unti-1819578279.

Neville, Robert Cummings. *Ultimates: Philosophical Theology*. Vol. 1. Albany: SUNY Press, 2014.

Nicolò, Anna Maria. "Playing." In *Playing and Reality Revisited: A New Look at Winnicott's Classic Work*, edited by Christian Seulin and Gennaro Saragnano, 21–44. London: Karmac Books, 2015.

Niebuhr, Richard R. *Schleiermacher on Christ and Religion*. London: SCM Press, 1965.

Nikkel, David. "Embodying Ultimate Concern." In *The Body and Ultimate Concern: Reflections on an Embodied Theology of Paul Tillich*, edited by Adam Pryor and Devan Stahl, 15–39. Macon, Ga.: Mercer University Press, 2018.

Nimmo, F., and D. McKenzie. "Volcanism and Tectonics on Venus." *Annual Review of Earth and Planetary Sciences* 26, no. 1 (May 1, 1998): 23–51.

Numan, A. P., V. C. Bennett, C. R. L. Friend, and M.D. Norman. "Waves and Weathering on the Early Earth: Geological Evidence for an Equable Terrestrial Climate at 3.7 GA." Lake Tahoe, Nevada, 2012. http://www.lpi.usra.edu/meetings/earlymars2012/pdf/7062.pdf.

Ojha, Lujendra, Mary Beth Wilhelm, Scott L. Murchie, Alfred S. McEwen, James J. Wray, Jennifer Hanley, Marion Massé, and Matt Chojnacki. "Spectral Evidence for Hydrated Salts in Recurring Slope Lineae on Mars." *Nature Geoscience* 8, no. 11 (November 2015): 829–32.

O'Neill, Craig, and Francis Nimmo. "The Role of Episodic Overturn in Generating the Surface Geology and Heat Flow on Enceladus." *Nature Geoscience* 3, no. 2 (February 2010): 88–91.

Ortlund, Gavin. "Image of Adam, Son of God: Genesis 5:3 and Luke 3:38 in Intercanonical Dialogue." *Journal of the Evangelical Theological Society* 57, no. 4 (December 2014): 673–88.

Pannenberg, Wolfhart. *Anthropologie in Theologischer Perspektive*. Göttingen: Vandenhoeck & Ruprecht, 1983.

Pascal, Robert. "Physicochemical Requirements Inferred for Chemical Self-Organization Hardly Support an Emergence of Life in the Deep Oceans of Icy Moons." *Astrobiology* 16, no. 5 (April 26, 2016): 328–34.

Patthoff, D. Alex, and Simon A. Kattenhorn. "A Fracture History on Enceladus Provides Evidence for a Global Ocean." *Geophysical Research Letters* 38, no. 18 (September 28, 2011): L18201.

Pavlov, Alexander A., James F. Kasting, Lisa L. Brown, Kathy A. Rages, and Richard Freedman. "Greenhouse Warming by Ch4 in the Atmosphere of Early Earth." *Journal of Geophysical Research: Planets* 105, no. E5 (May 25, 2000): 11981–90.

Peacocke, Arthur. *Theology for a Scientific Age: Being and Becoming—Natural, Divine, and Human*. Minneapolis: Fortress Press, 1993.

Pelikan, Jaroslav. *The Christian Tradition: A History of the Development of Doctrine*. 5 vols. Chicago: University of Chicago Press, 1971–1989.

Perry, M. E., B. D. Teolis, D. M. Hurley, B. A. Magee, J. H. Waite, T. G. Brockwell, R. S. Perryman, and R. L. McNutt Jr. "Cassini INMS Measurements of Enceladus Plume Density." *Icarus* 257 (September 1, 2015): 139–62.

Persson, Erik. "Philosophical Aspects of Astrobiology." In *The History and Philosophy of Astrobiology: Perspectives on Extraterrestrial Life and the Human Mind*, edited by David Dunér, Erik Persson, Gustav Holmberg, and Joel Parthmore, 29–48. Newcastle: Cambridge Scholars Publishing, 2013.

Petigura, Erik A., Andrew W. Howard, and Geoffrey W. Marcy. "Prevalence of Earth-Size Planets Orbiting Sun-like Stars." *Proceedings of the National Academy of Sciences* 110, no. 48 (November 26, 2013): 19273–78.

Pilcher, Carl. "Questing for Life: The Scientific Challenge." presented at the Center of Theological Inquiry Winter Symposium, Princeton, N.J., January 30, 2017.

Popkin, Gabriel. "What It Would Take to Reach the Stars." *Nature News* 542, no. 7639 (February 2, 2017): 20.

Postberg, F., J. Schmidt, J. Hillier, S. Kempf, and R. Srama. "A Salt-Water Reservoir as the Source of a Compositionally Stratified Plume on Enceladus." *Nature* 474, no. 7353 (June 30, 2011): 620–22.

Preus, H.D. "Dāmāh." Edited by G. Johannes Botterweck, Heinz-Josef Fabry, and Helmer Ringgren. Translated by Geoffrey W. Bromiley, David Green, and John T. Willis. *Theological Dictionary of the Old Testament*. Grand Rapids: Eerdmans, 2003.

Pryor, Adam. *Body of Christ Incarnate for You: Conceptualizing God's Desire for the Flesh.* Lanham, Md.: Lexington, 2016.

———. "Comparing Tillich and Rahner on Symbol: Evidencing the Modernist/Postmodernist Boundary." *Bulletin of the North American Paul Tillich Society* 37, no. 2 (Spring 2011): 23–38.

———. *The God Who Lives: Investigating the Emergence of Life and the Doctrine of God.* Eugene: Pickwick Publications, 2014.

———. "Intelligence, Non-Intelligence . . . Let's Call the Whole Thing Off." *Theology and Science* 16, no. 4 (October 2, 2018): 471–83.

———. "The Liminal as Kairos: A Cyborg Example for Theology and Science." In *The Many Voices of Liminality*, edited by Timothy Carson. Cambridge: Lutterworth Press, 2019.

Purdy, Jedediah. *After Nature: A Politics for the Anthropocene.* Cambridge: Harvard University Press, 2015.

"The Purpose of No-Till on the Plains." No-Till on the Plains: Agriculture Production Systems Modeling Nature, 2018. http://www.notill.org/about/purpose.

Rambaux, Nicolas, Julie C. Castillo-Rogez, James G. Williams, and Özgür Karatekin. "Librational Response of Enceladus." *Geophysical Research Letters* 37, no. 4 (February 1, 2010): L04202.

Raymond, Sean N., Rory Barnes, and Avi M. Mandell. "Observable Consequences of Planet Formation Models in Systems with Close-in Terrestrial Planets." *Monthly Notices of the Royal Astronomical Society* 384 (February 1, 2008): 663–74.

Rich, Adrienne. "Compulsory Heterosexuality and Lesbian Existence." In *The Lesbian and Gay Studies Reader*, edited by Henry Abelove, Michele Aina Barale, and David M. Halperin, 227–54. New York: Routledge, 1993.

Richards, Robert J. *The Romantic Conception of Life: Science and Philosophy in the Age of Goethe.* Chicago: University of Chicago Press, 2004.

Rigby, Kate. *Topographies of the Sacred: The Poetics of Place in European Romanticism.* Charlottesville: University of Virginia Press, 2004.

Rivera, Mayra. *Poetics of the Flesh*. Durham, N.C.: Duke University Press, 2015.

Rosenfield, P. L. "The Potential of Transdisciplinary Research for Sustaining and Extending Linkages between the Health and Social Sciences." *Social Science & Medicine* 35, no. 11 (December 1992): 1343–57.

Rubenstein, Mary-Jane. *Strange Wonder: The Closure of Metaphysics and the Opening of Awe*. New York: Columbia University Press, 2008.

Rubin, Henry S. "Phenomenology as a Method in Trans Studies." *GLQ* 4, no. 2 (1998): 145–58.

Rummel, John D., David W. Beaty, Melissa A. Jones, Corien Bakermans, Nadine G. Barlow, Penelope J. Boston, Vincent F. Chevrier, et al. "A New Analysis of Mars 'Special Regions': Findings of the Second Mepag Special Regions Science Analysis Group (SR-SAG2)." *Astrobiology* 14, no. 11 (November 2014): 887–968.

Russell, Robert J. *Cosmology—from Alpha to Omega: The Creative Mutual Interaction of Theology and Science*. Minneapolis: Fortress Press, 2008.

Sagan, Carl, and George Mullen. "Earth and Mars: Evolution of Atmospheres and Surface Temperatures." *Science* 177, no. 4043 (July 7, 1972): 52–56.

Sagan, Carl, W. Reid Thompson, Robert Carlson, Donald Gurnett, and Charles Hord. "A Search for Life on Earth from the Galileo Spacecraft." *Nature* 365, no. 6448 (October 21, 1993): 715–21.

Sartre, Jean-Paul. *Being and Nothingness*. Translated by Hazel E. Barnes. Reprint edition. New York: Washington Square Press, 1993.

Schleiermacher, Friedrich. *The Christian Faith*. Translated by H. R. Mackintosh and J. S. Stewart. Edinburgh: T & T Clark, 1960.

Schopf, J. William. *Earth's Earliest Biosphere: Its Origin and Evolution*. Princeton: Princeton University Press, 1983.

Schulze-Makuch, Dirk, and Louis Neal Irwin. *Life in the Universe: Expectations and Constraints*. Berlin: Springer, 2016.

Schutz, Alfred. "Multiple Realities, 1945." In *Collected Papers: The Problem of Social Reality*, 1:207–59. The Hague: Martinus Nijhoff, 1967.

Scott, Peter. *A Political Theology of Nature*. New York: Cambridge University Press, 2003.

Shults, F. LeRon. *Reforming Theological Anthropology: After the Philosophical Turn to Relationality*. Grand Rapids: Eerdmans, 2003.

Sideris, Lisa H. "Biosphere, Noosphere, and the Anthropocene: Earth's Perilous Prospects in a Cosmic Context." *Journal for the Study of Religion, Nature and Culture* 11, no. 4 (November 16, 2017): 399–419.

Smith, Bruce D., and Melinda A. Zeder. "The Onset of the Anthropocene." *Anthropocene* 4 (December 1, 2013): 8–13.

Smith, Eric, and Harold J. Morowitz. *The Origin and Nature of Life on Earth: The Emergence of the Fourth Geosphere*. New York: Cambridge University Press, 2016.

Southworth, John, Luigi Mancini, Nikku Madhusudhan, Paul Mollière, Simona Ciceri, and Thomas Henning. "Detection of the Atmosphere of the 1.6 $M_{\oplus}$ Exoplanet GJ 1132 b." *The Astronomical Journal* 153, no. 4 (2017): 191.

Spencer, John R., and Francis Nimmo. "Enceladus: An Active Ice World in the Saturn System." *Annual Review of Earth and Planetary Sciences* 41, no. 1 (2013): 693–717.

Spivak, Gayatri Chakravorty. *Death of a Discipline*. New York: Columbia University Press, 2003.

———. "Imperative to Re-Imagine the Planet." In *An Aesthetic Education in the Era of Globalization*, 335–50. Cambridge: Harvard University Press, 2013.

———. "'Planetarity' (Box 4, WELT)." *Paragraph* 38, no. 2 (July 2015): 290–92.

Squyres, S. W., and A. H. Knoll. *Sedimentary Geology at Meridiani Planum, Mars*. Amsterdam: Elsevier, 2005.

Stackhouse, Max L. "Civil Religion, Political Theology and Public Theology: What's the Difference?" *Political Theology* 5, no. 3 (July 2004): 275–93.

Steffen, Will, Wendy Broadgate, Lisa Deutsch, Owen Gaffney, and Cornelia Ludwig. "The Trajectory of the Anthropocene: The Great Acceleration." *The Anthropocene Review* 2, no. 1 (April 1, 2015): 81–98.

Steffen, Will, Jacques Grinevald, Paul Crutzen, and John McNeill. "The Anthropocene: Conceptual and Historical Perspectives." *Philosophical Transactions of the Royal Society of London A: Mathematical, Physical and Engineering Sciences* 369, no. 1938 (March 13, 2011): 842–67.

Steinbock, Anthony J. *Home and Beyond: Generative Phenomenology after Husserl*. Evanston: Northwestern University Press, 1995.

Stendebach, F. J. "Şelem." Edited by G. Johannes Botterweck, Heinz-Josef Fabry, and Helmer Ringgren. Translated by Douglas W. Stott. *Theological Dictionary of the Old Testament*. Grand Rapids: Eerdmans, 2003.

Stockton, Nick. "Elon Musk Announces His Plan to Colonize Mars and Save Us All." *Wired*, September 27, 2016. https://www.wired.com/2016/09/elon-musk-colonize-mars/.

Stokols, Daniel. "Toward a Science of Transdisciplinary Action Research." *American Journal of Community Psychology* 38, no. 1–2 (June 22, 2006): 63–77.

Szerszynski, Bronislaw. "Planetary Mobilities: Movement, Memory and Emergence in the Body of the Earth." *Mobilities* 11, no. 4 (August 7, 2016): 614–28.

———. "Viewing the Technosphere in an Interplanetary Light." *The Anthropocene Review*, October 19, 2016.

Taylor, Mark C. "End the University as We Know It." *The New York Times*, April 27, 2009, sec. Opinion. http://www.nytimes.com/2009/04/27/opinion/27taylor.html.

Taylor, Mark Lewis. *The Theological and the Political*. Minneapolis: Fortress Press, 2011.

Thatamanil, John. "Transreligious Theology as the Quest for Interreligious Wisdom : Open Theology." *Open Theology* 2, no. 1 (May 27, 2016): 354–62.

Thomas, Natalie, and Sumant Nigam. "Twentieth-Century Climate Change over Africa: Seasonal Hydroclimate Trends and Sahara Desert Expansion." *Journal of Climate* 31, no. 9 (March 28, 2018): 3349–70.

Thomas, P. C., R. Tajeddine, M. S. Tiscareno, J. A. Burns, J. Joseph, T. J. Loredo, P. Helfenstein, and C. Porco. "Enceladus's Measured Physical Libration Requires a Global Subsurface Ocean." *Icarus* 264 (January 2016): 37–47.

Thomas-Keprta, K. L., S. J. Clemett, D. S. McKay, E. K. Gibson, and S. J. Wentworth. "Origins of Magnetite Nanocrystals in Martian Meteorite ALH84001." *Geochimica et Cosmochimica Acta* 73, no. 21 (November 1, 2009): 6631–77.

Tillich, Paul. *The Courage to Be*. New Haven: Yale University Press, 2000.

———. *Dynamics of Faith*. New York: Harper-Collins, 2001.

———. "The Effects of Space Exploration on Man's Condition and Stature." In *The Future of Religions*, edited by Jerald C. Brauer, 39–51. New York: Harper & Row, 1966.

———. *Systematic Theology*. 3 vols. Chicago: University of Chicago Press, 1951–1963.

———. *Theology of Culture*. New York: Oxford University Press, 1959.

Toadvine, Ted. *Merleau-Ponty's Philosophy of Nature*. Evanston: Northwestern University Press, 2009.

Tobie, G., O. Čadek, and C. Sotin. "Solid Tidal Friction Above a Liquid Water Reservoir as the Origin of the South Pole Hotspot on Enceladus." *Icarus* 196 (August 1, 2008): 642–52.

Torres, Martha E. Sosa, Juan P. Saucedo-Vázquez, and Peter M. H. Kroneck. "The Magic of Dioxygen." In *Sustaining Life on Planet Earth: Metalloenzymes Mastering Dioxygen and Other Chewy Gases*, 1–12. Metal Ions in Life Sciences. Springer, Cham, 2015.

Turner, Ed. "Improbable Life." Presented at the Center of Theological Inquiry Colloquiums, Princeton, N.J., May 2, 2017.

Vainio, Olli-Pekka. "Imago Dei and Human Rationality." *Zygon: Journal of Religion & Science* 49, no. 1 (March 2014): 121–34.

Vilhauer, Monica. *Gadamer's Ethics of Play: Hermeneutics and the Other*. Lanham, Md.: Lexington Books, 2010.

"Vision and Mission." The Land Institute, 2018. https://landinstitute.org/about-us/vision-mission/.

Waite, J. Hunter, Michael R. Combi, Wing-Huen Ip, Thomas E. Cravens, Ralph L. McNutt, Wayne Kasprzak, Roger Yelle, et al. "Cassini Ion and Neutral Mass Spectrometer: Enceladus Plume Composition and Structure." *Science* 311, no. 5766 (March 10, 2006): 1419–22.

Waite Jr., J. H., W. S. Lewis, B. A. Magee, J. I. Lunine, W. B. McKinnon, C. R. Glein, O. Mousis, et al. "Liquid Water on Enceladus from Observations of Ammonia and 40ar in the Plume." *Nature* 460, no. 7254 (July 23, 2009): 487–90.

Walker, James C. G., P. B. Hays, and J. F. Kasting. "A Negative Feedback Mechanism for the Long-Term Stabilization of Earth's Surface Temperature." *Journal of Geophysical Research: Oceans* 86, no. C10 (October 20, 1981): 9776–82.

Wall, Mike. "Found! Potentially Earth-Like Planet at Proxima Centauri Is Closest Ever." Space.com, August 24, 2016. http://www.space.com/33834-discovery-of-planet-proxima-b.html.

———. "Life-Hunting Mission Would Bring Samples Back from Saturn Moon Enceladus." Space.com, September 21, 2015. http://www.space.com/30598-saturn-moon-enceladus-sample-return-mission.html.

Wang, Xunming, Fahu Chen, Jiawu Zhang, Yi Yang, Jijun Li, Eerdun Hasi, Caixia Zhang, and Dunsheng Xia. "Climate, Desertification, and the Rise and Collapse of China's Historical Dynasties." *Human Ecology* 38, no. 1 (2010): 157–72.

Way, M. J., Del Genio, Anthony D., Nancy Y. Kiang, Linda E. Sohl, David H. Grinspoon, Igor Aleinov, Maxwell Kelley, and Thomas Clune. "Was Venus the First Habitable World of Our Solar System?" *Geophysical Research Letters* 43, no. 16 (August 28, 2016): 8376–83.

Welch, Claude. *Protestant Thought in the Nineteenth Century*. 2 vols. Eugene: Wipf & Stock, 1972.

Welz, Claudia. "Imago Dei: References to the Invisible." *Studia Theologica* 65 (2011): 74–91.

Wenz, John. "Study Casts Doubt on Existence of a Potential 'Earth 2.0.'" *Scientific American*, April 9, 2018. https://www.scientificamerican.com/article/study-casts-doubt-on-existence-of-a-potential-earth-2-0/.

Wertheimer, Jeremy G., and Gregory Laughlin. "Are Proxima and Alpha Centauri Gravitationally Bound?" *The Astronomical Journal* 132, no. 5 (2006): 1995.

Westermann, Claus. *Genesis 1–11*. Translated by John J. Scullion. Hermeneia. Minneapolis: Fortress Press, 1994.

"Where Does It Hurt?" *On Being*. American Public Media, September 15, 2016. https://onbeing.org/programs/ruby-sales-where-does-it-hurt/.

Whittaker, Rebecca. "Learn About Spacesuits." NASA, June 5, 2013. http://www.nasa.gov/audience/foreducators/spacesuits/home/clickable_suit_nf.html.

Wickstrom, Shelley. "Introducing Shishmaref." Lecture, Convocation of the Association of Teaching Theologians, Minneapolis, July 31, 2018.

Wildberger, H. "Ṣelem—Image." In *Theological Lexicon of the Old Testament*, edited by Ernst Jenni and Claus Westermann, translated by Mark E. Biddle. Peabody, Mass.: Hendrickson Publishers, 1997.

Wildman, Wesley J. "Distributed Identity: Human Beings as Walking, Thinking, Ecologies in the Microbial World." In *Human Identity at the Intersection of Science, Technology, and Religion*, edited by Nancey Murphy and Christopher C. Knight. Burlington: Ashgate, 2010.

Willox, Anabelle. "Phenomenology, Embodiment and the Political Efficacy of Contingent Identity Claims." In *The Ashgate Research Companion to Queer Theory*, edited by Noreen Giffney and Michael O'Rourke, 1st edition, 95–110. Burlington, Vt.: Routledge, 2009.

Witze, Alexandra. "Earth-Sized Planet Around Nearby Star Is Astronomy Dream Come True." *Nature News* 536, no. 7617 (August 25, 2016): 381.

Woese, C. R., O. Kandler, and M. L. Wheelis. "Towards a Natural System of Organisms: Proposal for the Domains Archaea, Bacteria, and Eucarya." *Proceedings of the National Academy of Sciences of the United States of America* 87, no. 12 (June 1990): 4576–79.

Wolszczan, A., and D. A. Frail. "A Planetary System Around the Millisecond Pulsar Psr1257 + 12." *Nature* 355, no. 6356 (January 1992): 145–47.

Wray, James J., Scott L. Murchie, Janice L. Bishop, Bethany L. Ehlmann, Ralph E. Milliken, Mary Beth Wilhelm, Kimberly D. Seelos, and Matthew Chojnacki. "Orbital Evidence for More Widespread Carbonate-Bearing Rocks on Mars." *Journal of Geophysical Research: Planets* 121, no. 4 (April 1, 2016): 2015JE004972.

Wright, David K. "Humans as Agents in the Termination of the African Humid Period." *Frontiers in Earth Science* 5 (2017).

Wulf, Andrea. *The Invention of Nature: Alexander von Humboldt's New World*. New York: Knopf, 2015.

Yaqoob, Tahir. *Exoplanets and Alien Solar Systems*. Baltimore: New Earth Labs, 2011.

Zalasiewicz, J., M. Williams, A. G. Smith, T. L. Barry, A. L. Coe, P. R. Bown, P. Brenchley, et al. "Are We Now Living in the Anthropocene?" *GSA Today* 18, no. 2 (February 2008): 4–8.

Zeng, Ning, and Jinho Yoon. "Expansion of the World's Deserts Due to Vegetation-Albedo Feedback Under Global Warming." *Geophysical Research Letters* 36, no. 17 (September 1, 2009).

Žižek, Slavoj. *In Defense of Lost Causes*. London: Verso, 2017.

Zolotov, Mikhail Y. "An Oceanic Composition on Early and Today's Enceladus." *Geophysical Research Letters* 34, no. 23 (December 16, 2007): L23203.

Zolotov, Mikhail Y., and E. L. Shock. "An Abiotic Origin for Hydrocarbons in the Allan Hills 84001 Martian Meteorite Through Cooling of Magmatic and Impact-Generated Gases." *Meteoritics & Planetary Science* 35, no. 3 (May 2000): 629–38.

# Index

abiogenesis, 27, 30

abiotic: about, 3, 27, 30; avoiding 34–35, 41

abject, 104, 110

actualization, 13, 96

Adad-iti, 70

agential realism, 39–40, 213n15, 227n22

Ahmed, Sara, 157, 161, 234n27, 235nn30–31, 236n44

albedo effect, 190

Alexander of Hales, 4

ALH84001, 30, 33–34, 211n48

aliens: about, 4; and astrobiology, 13, 16–17, 34; sentient, 4, 36; tiny, 14, 30, 36, 199

Alpha Centauri a, 207n21

Alpha Centauri b, 207n21

alterity, 106–8, 110–14, 125–26, 140–41, 154, 156, 162, 195, 226n9

Amazonian, 28

Anselm of Canterbury, 4

Anthropocene: about, 87, 89–90, 95, 99–101, 104–5, 119, 122–23, 135, 138, 154, 221n1, 222n5, 226n2; and climate change, 85, 90, 117, 189, 197; as disorienting, 89, 92–93, 102–5, 124–25, 129, 140–41, 146; as eon, 116, 185, 228n41; epoch, 113–15, 123; and extinction, 230n62; and Holocene, 121; mature, 123, 126; proto-, 123–24, 231n70; as

symbol, 47–48, 88–91, 93, 100, 117, 126–27, 129–32, 136, 141, 144, 166, 178–79, 181, 198. *See also* planetarity; planetary

anthropocentrism 12, 98–99, 215n14

anthropology: about, 48; theological 8, 33, 53

anxiety, 5, 7, 13, 91, 93, 97–98, 124–25, 133–35, 141, 144, 162–63, 165–69, 175, 178, 183–84, 195, 222n13, 236n47

apologetic theology, 6, 48, 51, 62, 130–31

aqueous environment, 31

Arianism, 72

*Arrival*, 4

artful planet, 12–14, 123–24, 126, 136, 142–45, 147, 178–79, 184–85, 187, 199

astrobiology: about, 9–11, 13–17, 31–34, 38–41, 56–57, 179–80; and alterity, 113–14; and attunement, 3–5, 60–61; and conditions for life, 34–36; and cosmogony, 79–80; and context of engagement, 8, 13, 45, 47–48, 52, 55–57, 77–78, 84–91, 123–24, 127, 130, 132, 136–41, 178, 215n14, 220n45; and humanities, 5, 7, 12; institute, 4, 35; and similarity, 43–44; strategy document, 32, 34, 37. *See also* intra-action

astrometry, 18
Augustine, 63,67, 71–73, 77, 218n13,
    219nn22–23, 220n31

Bachelard, Gaston, 153
Barad, Karen, 39–40, 213n15, 227n22,
    237n74
Bauman, Whitney, 106, 109–10,
    227nn16,22
Bellah, Robert, 171
*Bestaunen*, 167
*Bewunderung*, 167
biogeochemical 14, 118–20, 122–23,
    126, 137, 140, 142–43, 178–79,
    192. *See also* geochemical;
    technobiogeochemical
biology: about, 15; evolutionary, 56; first
    principles, 40; systems, 11
biosignatures, 3, 35, 37, 41
biota: activity, 42; chemistry, 35; Mars,
    30, 33
de Biran, Maine, 4
black smokers, 26–27
Bohr, Niels, 213n15
Bonaventure, 4
bracketing, 236n44
Braudel, Fernand, 113, 228n33
Breakthrough Starshot Initiative, 22
Brown, Daniel, 18
Buber, Martin, 197
Burghardt, Gordan M., 175, 237n59

carnality. *See* flesh
Cassini, 25–27
cenozoic era, 114
Challenger, 1
Characterizing Exoplanets Satellite
    (CHEOPS), 19
chemoautotroph, 27, 31, 36, 38
chiasm, 111, 145, 150–56, 158–59, 161–
    62, 165, 167–68, 178, 186, 235n35.
    *See also* flesh
CHNOPS, 26, 35
chondritic rock, 27
christology: about, 9, 62, 65, 66, 68;
    *logos*, 65, 217n8
Clingerman, Forrest, 88, 222n5
Clinton, Bill, 30
Clipper Mission, 28

Clough, David, 12
co-habitation, 158
concomitance, 59
constitutive mutuality, 88–89
*Contact*, 4
continental zoning, 98
cosmic escapism, 45
cosmogony: about 48, 57–60, 79–80, 90,
    142, 216n28; biblical, 63
cost-benefit, 99, 225n38
cover cropping, 192
creative mutual interaction, 204n22
cross-disciplinary, 9–12, 16, 34–35, 56,
    204n22
crossed phenomenon, 154–55, 233n19
cyanobacteria, 120–21, 231n69
Cygnus, 17

Davies, Jeremy, 92–3, 112, 114–15, 122,
    142, 221n2
Deane-Drummond, Celia, 12
de-centering, 113, 125–26, 140–41, 186
deep time, 5, 14, 92, 103–5, 112–15,
    125–26, 129, 137, 140–41, 144, 147,
    154, 231n70
dehinscence. *See* flesh.
dermal metaphysics, 5, 110
Derrida, Jacques, 233n19
desertification, 189–92
disorientation, 91–93, 105, 112–13,
    124–26, 130, 134, 141, 146–47,
    161–63, 165–69, 173, 175, 178,
    183–84, 186–87, 198, 223nn13,14.
    *See also* orientation
dogmatic theology, 9
doppler effect, 20

Earth: age of, 92, 112; as artful planet,
    12, 14, 123–24, 178; beyond, 4,
    10, 15–16, 36–37; biosphere, 115,
    119; climate, 117–18; community,
    194–95; cousin, 18–19, 33; and
    ecosphere, 193; escaping, 42, 45; and
    habitat, 1–2, 137; history, 115–17,
    124–26, 129; and human activity,
    88–89; imaging, 3; and *imago Dei*,
    83, 141–43, 178; life on, 35, 46, 83;
    -like, 21–23, 25–27, 29–31, 42–44,
    164; location, 23, 41; and nature,

109; and origins of life, 27, 114; others, 110; -sized, 17–18, 21, 23; super-, 18–19; systems, 90, 113, 117–24, 126, 136–37, 140, 143

écart, 152, 161 *See also* flesh

ecosphere studies, 193, 195

*Eindeutung*, 144, 181

*Einfühlung*, 144, 181

elemental. *See* flesh

Enceladus, 16, 24–27, 31, 36, 38, 42–43, 208n29, 209n36

environmental reflexivity, 99–101, 103, 112, 114, 125, 225n42, 231n70

enzymatic reactions, 38

equiprimordial, 147

*Erlebnis*, 237n71

*Erstaunen*, 167

eschaton, 66

Europa, 25, 28

everydayness, 167, 171, 177–79, 183–84, 186, 237n71

evolution: 37, 115–16, 119; Darwinian, 36

existence, 6–9, 14, 16, 20, 25, 33, 37–41, 44–50, 52, 57–61, 67–80, 82–84, 86, 88, 90–93, 102–3, 106–8, 124–25, 132–39, 141–43, 146, 148, 154–56, 162–68, 176, 178–72, 184, 187, 197–99, 223n20, 227n16, 234n24

existential theology, 7–8

*ex nihilo*, 104, 158, 163

exobiology, 15

exoplanet; 15–20, 21–24, 29, 33, 36, 41, 136 163–66, 205n6, 206n14, 207n21, 229n48; system 205n6

Extravehicular Mobility Unit (EMU), 1–3

extremophiles, 26, 30, 209n36

Farley, Edward, 50

finite reality 49

finite ultimacy 49

flesh: about, 148–50, 163–64, 177–78, 179–87, 189–90, 233n9; and attention, 149; carnality, 148, 150–51; crossing in, 152; dehinscence, 151, 161; elemental, 147–49, 158–60; and failure, 160–63, 165–68, 172–73, 195; and incarnation, 65–66; as mutable,

153–54, 156, 168–69; and presence, 154–56, 197; reversibility, 150–54, 165. *See also* chiasm

fluvial, 28–9, 42, 102

Gadamer, Hans-Georg, 177, 237n68

Galileo spacecraft, 2–3, 28

geochemical, 37, 118–20, 122, 137. *See also* biogeochemical; technobiogeochemical

geophysical agency, 103–5, 119

Ghosh, Amitav, 92–93

Gibson, Everett, 33

Giri, A. K., 10

globalization, 106

god-consciousness, 71, 75–78, 81

Gopnik, Alison, 174, 176

grace, 75–76

Grinspoon, David, 11, 113–18, 120–26, 140, 229n47, 231n70

Gschwandtner, Christina, 222n5

habitablity: conceptualizing, 34–35, 37, 41; and EMU, 2; of exoplanets, 206n14, 229n48; instantaneous, 38, 41–42, 139; and living-systems, 5, 17, 32, 39–40, 57, 77, 113, 136; and Mars, 28, 43; planetary 38, 41–42, 117, 137, 139, 178

Hadean, 116

Hass, Lawrence, 232n2

Hawking, Stephen, 22

HD-142b, 163

Hesperian 28–29

heteronomy 179

Holocene, 87, 114, 121, 198

homo sapiens, 12, 116, 124, 137, 229n47

Huizinga, Johan, 173, 176

von Humboldt, Alexander, 4, 226n2

Van Huyssteen, Wentzel, 12, 53

hydrological: about 29; cycle, 118

hydrothermal, 26–27

hyper-individualism, 108

icy moons, 24, 29–30, 33, 42, 136, 209n36

IKAROS, 22

illusion, 171, 176

*imago Dei*: about, 7–9, 12–14, 62–63, 184–85; and alterity, 162; and astrobiology, 45, 47, 52, 56–57, 60–63, 78, 83–86, 138–43, 165, 178–79, 185–86, 192–93, 195, 220n45, 231n2; and cosmogony, 59–60, 79–80, 138, 216n28; ecological ethics, 189, 197, 214n14; Genesis, 58, 63–64, 67–68; human exceptionalism, 48, 53–55, 80, 138, 214n14; human perfection, 75–76; as image and likeness, 66–68, 70, 124, 218n9; *imago trinitate*, 63, 71–74, 219n23; *imitatio christi*, 65; individuality, 87, 103, 138; planetary, 141–43, 146, 154, 165–66, 179, 181, 184–85, 189; and recapitulation, 65–66, 217n8; and refraction, 81–91, 134, 138–39, 142, 184–87; as symbol, 13, 40, 48–49, 69–70, 77, 88–91, 102, 127, 130–32, 135–37, 144, 153, 186, 198; typology, 53–54; and ultimacy, 49, 62, 132, 137
incarnation: of flesh, 148, 152; of Christ 65–66
*Independence Day*, 4
*Ineinander*, 148, 169
*Ineinanderspiel*, 169
*Innerlichkeit*, 67, 78
interdisciplinary, 9–12, 17, 31–33, 45, 54, 215n17
interstellar travel, 22, 24, 42–43, 229n47
intra-action: about, 13–4, 39–40, 191, 213n15, 222n5, 227n22; and Anthropocene, 93, 104–5, 124, 129; and astrobiology, 33, 40, 47–48, 60–61, 80, 86, 89, 114, 136–37, 140, 178–81; and creation, 48, 84–85; and flesh, 180, 183, 195, 198, 237n74; of human being and nature, 88–91, 103, 109, 125; and *imago Dei*, 139–43, 154, 179, 192–93; of life and climate, 119; of living-systems and habitats, 16, 32, 44–45, 47–48, 56–57, 63, 77, 79, 83–85, 113, 116, 123, 127, 136–37, 139, 141–42, 154, 178–79, 199, 226n2; not interactive, 41; and planetarity, 107, 112, 116–17, 125–26, 137, 141; of self and world, 7, 145, 147–48; and thermophiles, 42

Irenaeus, 63, 65–67, 71, 78, 217n8, 218n12
*Irgendwiegewordensein*, 76

James Webb space telescope, 19
Jupiter, 20, 28

*kairos*, 149
Kalnay, Eugenia, 190
Kandler, Otto, 15
Kent, James, 96
Kepler-296e, 163
Kepler-452b, 16–20, 33, 42–43, 163–66
Kepler mission, 17–18, 205n6

Lederberg, Joshua, 15
Le Guin, Ursula, 146
Leonard, Annie, 128
Leopold, Aldo, 113–14
Leviathan, 58
Lewis, C. S., 93–95, 101, 223n20
Li, Yan, 190
Linzey, Andrew, 12
Liquid Cooling-and-Ventilation Garment (LCVG), 1–2
living-system: and astrobiology, 34–35, 37, 40–42, 56–57, 77, 80, 141; first principles, 44; general theory of 32–33, 36; and habitability 5, 7, 14, 16–17, 38–40, 45, 47, 56–57, 61, 63, 83–86, 89, 113, 116–17, 127, 136–37, 142, 154, 178–80, 198; as *imago Dei*, 87, 131, 138–42, 178–79, 181, 187, 192, 195; planetary, 120, 122, 136–37, 140–41, 178–79, 187; populations, 126; and solvent, 37, 43–44
Lost City, 26–27, 38, 43
Lubin, Phillip, 22

Macauley, David, 149
magnetosphere, 119
Manned Maneuvering Unit (MMU), 1
Marcionism, 65
Marion, Jean-Luc, 154, 233–34n19, 234n24
Marratto, Scott, 233n19
Mars, 16, 28–34, 36, 38, 42–43, 117–20, 136, 205n7, 211n45, 230n56
Marsh, George, 97, 224n32

materiality, 59, 101–2, 138, 148, 182
Mazis, Glen, 149
Melanchthon, Phillip, 82
Merleau-Ponty, Maurice, 148, 152,
    232nn2,5, 233nn14,19, 237n74
metabolism, 31, 36–38, 216n27
microbe, 4, 17, 27, 30–31, 33, 36–38, 43,
    110, 191, 199
monodisciplinary, 10
Muir, John, 96
multidisciplinary, 9–11
Musk, Elon, 43

Nancy, Jean-Luc, 233n19
nanoprobes, 22
naturalization, 104
*natura*, 109
nature: about, 13, 88–89, 101–2, 124–26,
    222n11; conservationist view of, 97–
    98; destabilizing, 227n16; ecological
    approach to, 98–99; as emergent pro-
    cess, 109–12; environmental imagi-
    nation, 93, 129, 223n18, 231n70;
    externality, 90; and human inter-
    ference, 99–101, 103; immanence
    of, 227n19; instrumentalized, 108;
    meaning of, 104–6; natural world,
    94; philology, 94, 223n20; politics
    of, 95, 224n26, 225n42; power over,
    141; preservationist view of, 96;
    providential view of, 95–96, 226n7;
    respect for, 197; romantic view of,
    96–97; as saturated phenomenon,
    222n5; temporality of, 113, 140;
    transcendentalism, 97
*Naturgemälde*, 226n2
neoplatonism, 4
Neville, Robert, 49
new materialism, 39, 104
Nicholas of Cusa, 4
Nigam, Sumant, 189
Noachian, 28–30, 42
nonhuman, 52

ontology: Being, 149; Buber, 197; cat-
    egories, 110; discrete subjects, 44;
    elemental, 148, 158, 177, 179, 182,
    232n2; entanglement, 39; frames of
    reference, 44, 102, 177; and fusion

of horizons, 237n68; language, 5, 13,
    179, 181; Merleau-Ponty, 233n14;
    penumbral, 153; and the primor-
    dial, 8, 39–40, 153, 156, 180, 222n5;
    reason, 222n11; relationality, 55;
    shift, 88; substantialist, 149; units,
    5, 33, 39–40, 153–54, 156, 213n15;
    as untenable, 102; vocabulary, 112;
    wholeness, 237n71. *See also* flesh
Oort cloud, 23
organism: -environment interrelation-
    ship, 98–99; individual, 17, 56,
    63, 84, 122, 179; and life, 40, 42,
    120; micro-, 26, 31; multicellular,
    121; photosynthetic, 3; terrestrial,
    211n45
orientation: -around, 157–60, 162–66,
    182–83, 193, 195; and harmony, 161;
    indirect, 235n35; and intentionality,
    235n30; and performativity, 235n31;
    and presence, 147; queer, 166,
    188; re-, 193, 195, 235n37; sexual,
    156–57; -toward, 157–60, 162–65,
    172–73, 181–83, 193, 195; trajectory,
    158. *See also* disorientation
oxygenation, 120–21, 231n69

Pauline letters, 66, 218n9
performative: agency, 227n22; and
    constructivism, 104; dimension,
    173; disruption, 235n37; habits, 111;
    repetition, 109; tendencies, 157,
    235n31
Phanerozoic, 114, 116–17, 123, 143
photodissociate, 3
photometer, 17
Pilcher, Carl, 4, 35–37, 43
planetarity, 5, 8, 14, 105–12, 114, 119,
    125–26, 129, 132, 136, 140–41, 144,
    156, 165
planetary: being, 195; bodies, 15, 35;
    catastrophe, 117; cephalization,
    123; control, 45, 114, 116; creatures,
    5, 14, 105, 107–8, 111–12, 125–26,
    144, 147, 154; crisis, 13; flesh, 187;
    flourishing, 111, 144; flows, 14, 109,
    137, 143; formation, 21; habitability,
    38–39, 41–42, 117, 119, 137, 139,
    178; history, 12, 29, 31, 115–17, 124,

planetary (*continued*)
    228n41; identity, 110–11, 126; *imago Dei*, 139–40, 181, 189, 192–93, 198; knowing, 107–8, 110, 126; quality of life, 40, 136; others, 6, 52, 84, 11; phenomenon, 14, 37, 141–42,146; probe, 3; problems, 129; resources, 128; responsibility, 143; scale, 121, 140; scientist, 16, 19; sustainability, 12; thinking, 13; time scale, 113, 115, 124; usury, 229n47
Pleistocene, 114, 117
pluvial, 28
*poiesis*, 173, 176–77, 181, 185, 222n11
pollution, 87, 98–100, 188
polymers, 38
praxis, 10, 177, 185
Primary Life-Support System, 1
primordial unit. *See* ontology
*prius*, 59, 80
Proterozoic, 116
providential. *See* nature
Proxima Centauri b, 16, 19–24, 42–43, 206n14, 207nn19,21
Proxima Centauri c, 207n20
proximate: becoming, 158; concerns, 50; desires, 147, 235n37; experience, 160; gardener, 53; intra-action, 194–95; Kepler-452b, 163; loss of, 165, 183; objects, 157–64
Pseudo-Dionysius, 4
public theology, 6, 13, 132, 144
Pulsar B1257+12, 15
Purdy, Jedediah, 93–96, 99–102, 223nn18,23; 224n26

queer: about, 235n42; disorientation devices, 166, 168; effects, 173; exoplanet, 166; persons, 234n27; phenomenology, 156–57, 182, 234n27, 236n44; theory, 157, 235n31; exoplanet. *See also* orientation

radial velocity, 18, 20–21
radiative-convective climate modeling, 205n7
radioactive, 87
raw material, 94

recurring slope lineae (RSL), 29, 211n45
relata, 39
*res cogitan*, 148
*res extensa*, 148
reversibility. *See* flesh
Rivera, Mayra, 223n19
romanticism. *See* nature
Rosenfield, P. L., 9
Ross 128, 207n19
Ross-128b, 24, 207n19
Rubenstein, Mary, 167, 236n47
Russell, Robert, 204n22

Sagan, Carl, 3, 212n7
salvation: and creation, 65–66; history, 63, 75, 78–79; soteriology, 53, 65
Sapiezoic, 105, 116–17, 123, 126, 185, 229n47
Schleiermacher, Friedrich, 63, 71, 74–78,
Schutz, Alfred, 170
Seneca the Younger, 188
*Septuagint*, 217n4, 218n9
serpentinization, 26–27, 42, 209nn36,37
SETI, 36
Settlers of Catan, 73
Shishmaref, 196–98
sin: antithesis to grace, 75–76; and anxiety, 223n13; doctrine of, 53; and *imago Dei*, 67
*Solaris*, 4
soteriology. *See* salvation
*The Sparrow*, 4
species of alterity. *See* alterity
species of eternity, 107
spectral lines, 20
spectrograph, 20
spectroscopy, 20
Spivak, Gayatri Chakravorty, 105–8
star: *ad astra*, 188; G-type, 21, 23–24, 164; G2-type, 42; K-type, 206n14; M-type, 21–24, 206n14, 207n19; observations, 41, 205n6, 206n13; red dwarf, 21; spectrum of light, 20; system, 207n21
*Staunen*, 167
Steele, Andrew, 33
sub-Neptune, 19

super-Earth, 18–19
symbol: about, 49–51, 72, 89–90, 102, 130–32, 136; and Anthropocene, 13, 88, 90, 93–94, 99–100, 103, 129, 132, 166, 222n5; aspirational, 12; corroborative, 91, 134–35; doctrinal, 13, 51–52, 130, 133; and *imago Dei*, 48, 56–57, 60, 63, 68–70, 74, 77, 81, 84, 90, 103, 131–32, 138–39, 144, 215n14, 219n17, 220n45; for intra-action, 13, 91, 129

technical reason, 91, 222n11
technobiogeochemical, 12, 14, 122–23, 142–44, 146, 178–79, 181, 185, 189, 192, 199. *See also* biogeochemical; geochemical
theology and science, 52–54, 60, 86
thermodynamics, 3, 27, 212n5
thermophiles, 42
Thomas, Natalie, 189
Tillich, Paul, 49, 67, 214n2, 216n24, 222n11
Titan, 24, 37
transcendentalism. *See* nature
transdisciplinary, 9–13, 33, 43–44, 57, 143, 215n17
Transiting Exoplanet Survey Satellite (TESS), 23
transversal, 12, 53

Trappist-1, 23–24

von Uexkull, Jakob Johann, 4
ultimacy, 6, 9, 13, 49–53, 56–57, 59, 62, 70, 72, 77, 80–81, 88, 90–91, 124, 130–37, 144–45
ultimate: concern, 4, 50, 132; existence, 90–91; fulfillment, 134; as God, 79; reality, 49

Vainio, Olli-Pekka, 54
Venus, 117–20, 205n7
*Verwunderung*, 167
Victorines, 4
Virgil, 188
Voyager 2, 24

*Weltmöglichkeit*, 152–53
Wheelis, Mark, 15
Wickstrom, Shelley, 196–97
Wildberger, H., 216n1
Woese, Carl, 15

xenobiology, 15

Yoon, Jinho, 190

*Zeitgeist*, 99
Zeng, Ning, 190

**Adam Pryor** is Associate Professor of Religion and Dean of Academic Affairs at Bethany College. He is the author of two other books: *Body of Christ Incarnate for You: Conceptualizing God's Desire for the Flesh* (Lexington, 2016) and *The God Who Lives: Investigating the Emergence of Life and the Doctrine of God* (Pickwick, 2014).

# gROUNDWORKS |

## ECOLOGICAL ISSUES IN PHILOSOPHY AND THEOLOGY

Forrest Clingerman and Brian Treanor, series editors

*Interpreting Nature: The Emerging Field of Environmental Hermeneutics*
   Forrest Clingerman, Brian Treanor, Martin Drenthen,
   and David Utsler, eds.

*The Noetics of Nature: Environmental Philosophy
and the Holy Beauty of the Visible*
   Bruce V. Foltz

*Environmental Aesthetics: Crossing Divides
and Breaking Ground*
   Martin Drenthen and Jozef Keulartz, eds.

*The Logos of the Living World: Merleau-Ponty, Animals, and Language*
   Louise Westling

*Being-in-Creation: Human Responsibility in an Endangered World*
   Brian Treanor, Bruce Ellis Benson, and Norman Wirzba, eds.

*Wilderness in America: Philosophical Writings.
Edited by David W. Rodick*
   Henry Bugbee

*Eco-Deconstruction: Derrida and Environmental Philosophy*
Matthias Fritsch, Philippe Lynes, and David Wood, eds.

*Animality: A Theological Reconsideration*
Eric Daryl Meyer,

*Reoccupy Earth: Notes toward an Other Beginning*
David Wood

*Living with Tiny Aliens: The Image of God for the Anthropocene*
Adam Pryor